Current Topics in Microbiology and Immunology

Volume 435

The reviews series *Current Topics in Microbiology and Immunology* publishes cutting-edge syntheses of the latest advances in molecular immunology, medical microbiology, virology and biotechnology. Each volume of the series highlights a selected timely topic, is curated by a dedicated expert in the respective field, and contains a wealth of information on the featured subject by combining fundamental knowledge with latest research results in a unique manner.

2020 Impact Factor: 4.291, 5-Year Impact Factor: 5.110

2020 Eigenfaktor Score: 0.00667, Article Influence Score: 1.480

2020 Cite Score: 7.7, h5-Index: 38

More information about this series at https://link.springer.com/bookseries/82

Mark Heise
Editor

Chikungunya Virus

Responsible Series Editor: Michael Oldstone

Springer

Editor
Mark Heise
Department of Genetics
The Department of Microbiology
and Immunology
University of North Carolina
Chapel Hill, NC, USA

ISSN 0070-217X ISSN 2196-9965 (electronic)
Current Topics in Microbiology and Immunology
ISBN 978-3-030-90609-2 ISBN 978-3-030-90610-8 (eBook)
https://doi.org/10.1007/978-3-030-90610-8

This Springer imprint is published by the registered company Springer Nature Switzerland AG
The registered company address is: Gewerbestrasse 11, 6330 Cham, Switzerland

Preface

Chikungunya virus (CHIKV) is a mosquito-borne alphavirus in the family *Togaviridae* that is associated with large scale outbreaks of debilitating polyarthralgia [1]. While CHIKV has historically been associated with localized disease outbreaks in Africa and Southeast Asia, over the past 20 years, factors such as increased global travel, an expansion in the geographic distribution of transmission competent mosquito vectors, and viral adaptation to new mosquito vectors has led to several major CHIKV outbreaks [2–5]. These include largescale outbreaks in the Indian Ocean region and the Americas that illustrate CHIKV's capacity as a global disease threat [4].

Largescale CHIKV outbreaks, which can rapidly affect a significant fraction of the population [6], occur in an urban epidemic transmission cycle involving direct human to mosquito transmission [7]. The explosive nature of these outbreaks is driven both by the virus's ability to generate high levels of serum viremia in humans and efficient transmission by Aedes species mosquito vectors [8-10]. *A. aegypti* mosquitoes are the main mediator of epidemic CHIKV transmission, however, the virus has shown the ability to adapt to *A. albopictus* mosquitoes [11], an aggressive mosquito species with a broader geographic range than *A. aegypti* [12], that has facilitated a further expansion in CHIKV's geographic range [5].

CHIKV's ability to be efficiently transmitted and its high symptomatic attack rate can result in the rapid development of high local caseloads that can quickly overwhelm public health systems and drive the transmission of the virus into new areas [1]. CHIKV causes a wide range of disease signs, including fever, rash, nausea, and severe polyarthralgia and myalgia (Reviewed in [1, 6]). While CHIKV-associated mortality rates are generally low, the severe polyarthralgia and myalgia caused by CHIKV is often debilitating, and outbreaks of CHIKV-induced arthralgia can significantly impact quality of life and overwhelm local health resources [1, 4, 6]. Furthermore, virus-induced arthralgia can persist for months to years in a subset of infected individuals, leading to long-term negative impacts on quality of life [1, 4].

The growing importance of CHIKV as an emerging disease threat has led to a significant increase in research into CHIKV biology. This includes investigation into the fundamental biology of CHIKV replication and the study of how interactions between viral components and host factors regulate CHIKV replication and pathogenesis. There has also been a major effort devoted to developing models of CHIKV disease and using these model systems to identify viral and host factors that contribute to virus induced disease, while also characterizing aspects of the host immune response that limit CHIKV replication and protect from virus induced disease. Lastly, a significant effort has been focused on the development of new therapeutics or vaccines for the treatment or prevention of CHIKV-induced disease. The purpose of the reviews in this issue is to highlight key aspects of this research in areas related to viral replication and disease pathogenesis, as well as antiviral drug and vaccine development for the treatment and/or prevention of CHIKV-induced disease.

As noted above, since CHIKV re-emerged in the mid 2000's, the alphavirus research field has made significant strides in understanding the biology of the CHIKV replication. In Chapter "Molecular Virology of Chikungunya Virus," Frolov and Frolova provide an in depth overview of alphavirus RNA synthesis. This includes a comprehensive summary of the role that viral RNA elements and virally encoded nonstructural proteins play in the regulation of viral RNA synthesis. They also discuss key advances in our understating of how host proteins are coopted to promote CHIKV replication, a relatively new and rapidly moving area of alphavirus research that promises to provide new insights into the regulation of alphavirus replication and cell tropism, while also identifying potential targets for therapeutic intervention.

Much of our understanding of CHIKV virology and pathogenesis has been driven by the availability of a robust toolkit of molecular virology reagents, as well as *in vitro* and *in vivo* systems for studying viral replication, transmission, and disease that has been developed since CHIKV re-emerged as a global disease threat. In their chapter titled "Understanding Molecular Pathogenesis with Chikungunya Virus Research Tools," Carissimo and Ng discuss how research tools such a viral reverse genetics systems and reporter viruses, as well as cell culture systems, and mosquito and vertebrate animal models of CHIKV replication and disease have provided important insights into a wide range of topics related to CHIKV replication, virus/host interactions, immunity, and disease pathogenesis.

While CHIKV infection is generally associated with acute febrile illness and polyarthralgia, a significant fraction of infected individuals develop chronic disease signs, including relapsing arthralgia, that can last for month to years and significantly impact long-term quality of life [1]. Although the long-term disease manifestations were observed during the earliest documented CHIKV outbreaks [13], and have long been considered to be a significant aspect of CHIKV infection and disease [1], the massive scale of more recent CHIKV outbreaks has led to a significant increase in the overall number of people suffering from chronic CHIKV disease [14]. This has led to a renewed interest in understanding the pathogenesis of long-term CHIKV-associated poly-arthralgia. In their chapter titled "Chronic Chikungunya Virus Disease," McCarthy, Davenport, and Morrison provide an overview of the history of chronic CHIKV disease, and discuss the specific symptoms and risk factors associated with the development of post-acute CHIKV disease. They also discuss the potential role of immune and viral factors in the pathogenesis of chronic CHIKV disease, while providing an overview of strategies for treating persistent CHIKV-induced disease.

CHIKV poses a significant re-emerging disease threat that is capable of causing significant human suffering and economic disruption, however as of this writing, there are no approved virus-specific antiviral therapies or vaccines for the treatment or prevention of CHIKV. Victor DeFilippis, in his chapter titled "Chikungunya Virus Vaccines: Platforms, Progress, and Challenges," provides an overview of CHIKV pathogenesis and correlates of immune protection, while discussing the wide variety of vaccine platforms that have been investigated or are in development as CHIKV vaccines. While existing CHIKV disease treatments are largely limited to supportive therapy and pain management, the development of antiviral therapies for treating CHIKV-induced disease has also been an active area of scientific investigation since the virus's re-emergence in the mid-2000's. Haese, Powers, and Streblow, in their chapter titled "Small Molecule Inhibitors Targeting Chikungunya Virus," provide an overview of CHIKV biology and highlight specific stages in the viral replication cycle that can be targeted by direct acting antivirals, while discussing key advances in these areas. They also discuss host pathways that are essential for the viral replication cycle, and highlight specific strategies for targeting these pathways to develop potential broad spectrum inhibitors of CHIKV and other alphaviruses.

In summary, the chapters in this issue provide in depth reviews on key aspects of CHIKV biology and disease research, while highlighting specific areas related to vaccine or therapeutic development against this important re-emerging pathogen. Furthermore, the authors highlight areas within the rapidly moving CHIKV research field where additional work is needed both to understand CHIKV biology and to promote the development of safe and effective vaccines or therapeutics for the treatment or prevention of CHIKV-induced disease.

Chapel Hill, USA Mark Heise

Literature Cited

1. Burt FJ, Rolph MS, Rulli NE, Mahalingam S, Heise MT (2012) Chikungunya: a re-emerging virus. Lancet 379(9816):662–671. Epub 2011/11/22. doi: S0140-6736(11)60281-X [pii]. 10.1016/S0140-6736(11)60281-X. PubMed PMID: 22100854

2. Powers AM, Logue CH (2007) Changing patterns of chikungunya virus: re-emergence of a zoonotic arbovirus. J Gen Virol 88(Pt 9):2363–2377. Epub 2007/08/19. 10.1099/vir.0.82858-0. PubMed PMID: 17698645
3. Petersen LR, Powers AM (2016) Chikungunya: epidemiology. F1000Res 5. Epub 2016/02/27. 10.12688/f1000research.7171.1. PubMed PMID: 26918158; PubMed Central PMCID: PMCPMC4754000
4. Morrison TE (2014) Reemergence of chikungunya virus. J Virol 88 (20):11644–7. Epub 2014/08/01. 10.1128/JVI.01432-14. PubMed PMID: 25078691; PubMed Central PMCID: PMCPMC4178719
5. Townson H, Nathan MB (2008) Resurgence of chikungunya. Trans R Soc Trop Med Hyg 102(4):308–9. Epub 2008/01/08. S0035-9203(07)00404-X [pii]. 10.1016/j.trstmh.2007.11.013. PubMed PMID: 18178232
6. Powers AM (2010) Chikungunya. Clin Lab Med 30(1):209–19. Epub 2010/06/02. S0272-2712(09)00120-6 [pii]. 10.1016/j.cll.2009.10.003. PubMed PMID: 20513548
7. Silva LA, Dermody TS (2017) Chikungunya virus: epidemiology, replication, disease mechanisms, and prospective intervention strategies. J Clin Invest 127 (3):737–49. Epub 2017/03/02. 10.1172/JCI84417. PubMed PMID: 28248203; PubMed Central PMCID: PMCPMC5330729
8. Ledermann JP, Borland EM, Powers AM (2017) Minimum infectious dose for chikungunya virus in Aedes aegypti and Ae. albopictus mosquitoes. Rev Panam Salud Publica 41:e65. Epub 2017/09/14. PubMed PMID: 28902278; PubMed Central PMCID: PMCPMC6612711
9. Dubrulle M, Mousson L, Moutailler S, Vazeille M, Failloux AB (2009) Chikungunya virus and Aedes mosquitoes: saliva is infectious as soon as two days after oral infection. PLoS ONE 4(6):e5895. Epub 2009/06/13. 10.1371/journal.pone.0005895. PubMed PMID: 19521520; PubMed Central PMCID: PMC2690823
10. Staples JE, Breiman RF, Powers AM (2009) Chikungunya fever: an epidemiological review of a re-emerging infectious disease. Clin Infect Dis 49(6):942–8. Epub 2009/08/12. 10.1086/605496. PubMed PMID: 19663604
11. Tsetsarkin KA, Vanlandingham DL, McGee CE, Higgs S (2007) A single mutation in chikungunya virus affects vector specificity and epidemic potential. PLoS Pathog 3(12):e201. Epub 2007/12/12. 07-PLPA-RA-0664 [pii]. 10.1371/journal.ppat.0030201. PubMed PMID: 18069894; PubMed Central PMCID: PMC2134949
12. Paupy C, Delatte H, Bagny L, Corbel V, Fontenille D (2009) Aedes albopictus, an arbovirus vector: from the darkness to the light. Microbes Infect 11(14–15):1177–85. Epub 2009/05/20. S1286-4579(09)00105-1 [pii]. 10.1016/j.micinf.2009.05.005. PubMed PMID: 19450706
13. Robinson MC (1955) An epidemic of virus disease in Southern Province, Tanganyika Territory, in 1952–53. I. Clinical features. Trans R Soc Trop Med Hyg 49(1):28–32. Epub 1955/01/01. 10.1016/0035-9203(55)90080-8. PubMed PMID: 14373834

14. Paixao ES, Rodrigues LC, Costa M, Itaparica M, Barreto F, Gerardin P, et al (2018) Chikungunya chronic disease: a systematic review and meta-analysis. Trans R Soc Trop Med Hyg 112(7):301–16. Epub 2018/07/15. 10.1093/trstmh/try063. PubMed PMID: 30007303

Contents

Molecular Virology of Chikungunya Virus

I. Frolov and E. I. Frolova

Contents

Abstract Chikungunya virus (CHIKV) was discovered more than six decades ago, but has remained poorly investigated. However, after a recent outbreak of CHIK fever in both hemispheres and viral adaptation to new species of mosquitoes, it has attracted a lot of attention. The currently available experimental data suggest that molecular mechanisms of CHIKV replication in vertebrate and mosquito cells are similar to those of other New and Old World alphaviruses. However, this virus exhibits a number of unique characteristics that distinguish it from the other, better studied members of the alphavirus genus. This review is an attempt to summarize the data accumulated thus far regarding the molecular mechanisms of alphavirus RNA replication and interaction with host cells. Emphasis was placed on demonstrating the distinct features of CHIKV in utilizing host factors to build replication complexes and modify the intracellular environment for efficient viral replication and inhibition of the innate immune response. The available data suggest that our

I. Frolov (✉) · E. I. Frolova
Department of Microbiology, University of Alabama at Birmingham, 1720 2nd Ave South, BBRB373/Box 3, 35294-2170 Birmingham, AL, USA
e-mail: ivfrolov@uab.edu

Current Topics in Microbiology and Immunology (2022) 435: 1–31
https://doi.org/10.1007/82_2018_146

Published online: 01 January 2019

knowledge about alphavirus replication contains numerous gaps that potentially hamper the development of new therapeutic means against CHIKV and other pathogenic alphaviruses.

1 Introduction

Chikungunya virus (CHIKV) was first discovered in the 1952–1953 Tanzania outbreak (Robinson 1955). For decades, it has not been investigated as intensively as other alphaviruses, such as Sindbis virus (SINV), Semliki Forest virus (SFV), or Venezuelan equine encephalitis virus (VEEV). However, after 2005, CHIKV has attracted additional attention because of its transmission to new areas all over the globe, causing millions of cases of highly debilitating disease, which frequently lead to chronic arthralgia (Weaver and Forrester 2015; Weaver and Lecuit 2015a, b; McSweegan et al. 2015; Tsetsarkin et al. 2007). Since 2013, it has been circulating in Southern and Central Americas and Caribbean Islands (Langsjoen et al. 2016), but so far is not continuously present in the USA, albeit there have already been numerous imported cases (Nappe et al. 2016).

To date, the molecular mechanism of CHIKV replication and virus–host interactions remains insufficiently understood. However, CHIKV shares many aspects of its biology with other alphaviruses. Thus, a large part of the accumulated knowledge about mechanism of replication and virus–host interactions generated on other alphaviruses, particularly those circulating in the Old World (OW), can be extrapolated to CHIKV. Because of their less pathogenic phenotype, other OW alphaviruses, such as Sindbis virus (SINV) and Semliki Forest virus (SFV), were serving for decades as widely accepted models in the studies of molecular mechanism of replication and virus–host interactions (Rupp et al. 2015; Strauss and Strauss 1994).

2 Chikungunya Virus Genome and *Cis*-Acting Sequence Elements

As in other alphaviruses, the CHIKV genome is represented by a single-stranded RNA of positive polarity ~12 kb in length (Khan et al. 2002). This genomic RNA (G RNA) mimics the structure of cellular mRNAs in that it has a Cap at the 5′ terminus and a poly(A) tail at the 3′ terminus (Fig. 1). However, in contrast to cellular RNA templates, (i) the first nucleotide in the alphavirus genomes is not methylated, and the resulting Cap structure is termed Cap(0). (ii) Polyadenylated alphavirus-specific RNAs are synthesized in the cytoplasm of infected cells without using nuclear mRNA polyadenylating machinery. (iii) Alphavirus G RNAs contain a number of nucleotide sequences, which either function in the RNA replication and

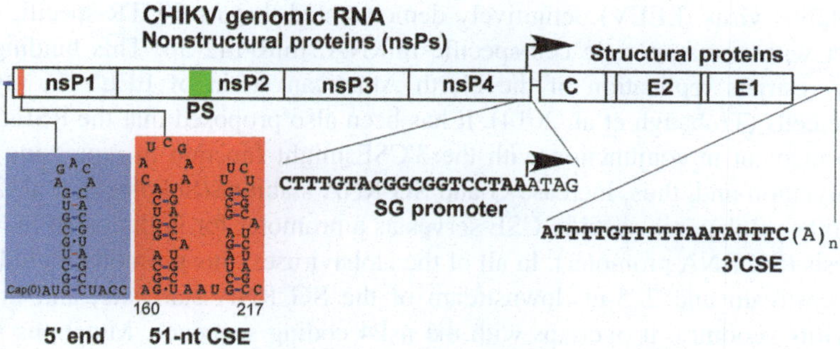

Fig. 1 Schematic presentation of CHIKV genomic RNA and *cis*-acting sequence elements. PS indicates packaging signal

transcription of subgenomic RNA (SG RNA) or determine the specificity of RNA packaging into released virions. They are termed the *cis*-acting sequence elements (CSEs) (Fig. 1).

The first CSE is located at the 3′ end of G RNA immediately upstream of the poly(A) tail (Kuhn et al. 1990). At early times post-infection, during amplification of viral genomes, this 19-nt-long RNA sequence functions as a critical element of the promoter for negative-strand RNA synthesis. Mutations in the 3′ CSE have a deleterious effect on the ability of in vitro-synthesized SINV RNAs to initiate replication upon their transfection into vertebrate cells (Frolov et al. 2001). However, in a few cells, these defective viral RNAs acquired heterologous AU-rich sequences, which were again capable of functioning as promoters in RNA replication. These newly developed RNAs initiated spreading infection, but the released viruses replicated less efficiently than the parental SINV having a natural CSE in the genome (Raju et al. 1999). The 3′ CSE is likely not the only element required for efficient negative-strand synthesis, because i) so far there is no experimental evidence that viral subgenomic RNA (SG RNA), which contains the same 3′ CSE, can serve as a template for negative-strand RNA synthesis, and ii) modifications in the 5′ end of SINV G RNA can also make it a very inefficient template for the negative-strand RNA synthesis in cell-free in vitro replication systems. This suggests a cooperative function of the 5 same 3′CSE and 3′ genome ends in initiating negative-strand RNA synthesis (Frolov et al. 2001).

The 3′ untranslated region (3′UTR) of CHIKV G RNA that is located upstream of the 3′CSE contains a number of repeating nucleotide sequences termed repeating sequence elements (RSEs) (Ou et al. 1982). Their function in CHIKV replication still remains to be understood. However, the accumulating experimental data suggest that CHIKV-specific 3′UTRs, and RSEs in particular, play critical roles in viral interaction with mosquito cell-specific factors and adaptation to mosquito vectors (Chen et al. 2013). RSEs have been identified in 3′UTRs of essentially all the alphaviruses studied to date, but they are very diverse. Extended deletions or replacements in the 3′UTRs usually have no deleterious effect on viral replication in cultured cells (Kuhn et al. 1990, 1991). However, a study using eastern equine encephalitis virus (EEEV) definitively demonstrated that its 3′UTR-specific RSEs interact with hematopoietic cell-specific miRNA, miR-142-3p. This binding efficiently restricts replication of the North American strain of EEEV in myeloid lineage cells (Trobaugh et al. 2014). It has been also proposed that the RSEs and a U-rich element in conjunction with the 3′CSE might function in repressing RNA deadenylation and, thus, increase G and SG RNA stability (Sokoloski et al. 2010).

Another alphavirus-specific CSE serves as a promoter for initiation of SG RNA synthesis (SG RNA promoter). In all of the alphaviruses, this promoter includes the 19-nt upstream and 2–5-nt downstream of the SG RNA start site, and in most alphavirus genomes, it overlaps with the nsP4 coding sequence. Mutations in this sequence that do not affect the encoded amino acid sequence of nsP4 either deleteriously affect efficiency of SG RNA synthesis (Grakoui et al. 1989) or completely abrogate transcription of CHIKV- and VEEV-specific SG RNAs (Volkova et al. 2008; Plante et al. 2011; Kim et al. 2011a). The SG RNA promoter acts on the negative-strand RNA intermediate, but the exact mechanism of its interaction with the viral replication complex remains to be understood.

The 5′UTRs of alphavirus genomes do not demonstrate sequence conservation, with the exception of having an AU as the first two nucleotides. Mutations in these positions abrogate G RNA replication (Niesters and Strauss 1990; Kulasegaran-Shylini et al. 2009a). The very 5′-terminal 23–45 nt of G RNAs (25 nt for CHIKV) exhibit a conservative feature as they are predicted to fold into stem-loop structures (Reynaud et al. 2015). The existence of these stem-loops was confirmed by enzymatic and structural analysis in some alphavirus genomes (Kulasegaran-Shylini et al. 2009b). Presence of the stem-loop structures at the very 5′ termini likely makes translation of G RNAs less efficient. In the case of VEEV, a positive effect of stem-destabilizing mutations on G RNA translation has been experimentally detected (Kulasegaran-Shylini et al. 2009b). On the other hand, the involvement of the first nucleotides in stem formation makes Cap(0) less accessible for binding by cellular interferon-induced protein with tetratricopeptide repeats, IFIT1 (Reynaud et al. 2015; Hyde et al. 2014). This protein specifically interacts with Cap (0) and blocks its interaction with the cellular translation initiation factor eIF4E. However, the IFIT1–Cap(0) interaction also requires 3–4 nt that follow Cap to be not involved into base pairing (Kumar et al. 2014). Thus, the 5′ RNA stems protect alphavirus genomes from being sensed by IFIT1 and, thus, excluded from translation. This makes translation of alphavirus G RNA more resistant to the antiviral state induced by type I IFN treatment. A G-to-A point mutation of nt3 in the CHIKV genome has deleterious effect on viral replication in cell lines stably overexpressing IFIT1, but not in the cells with baseline low levels of expression of this protein (Reynaud et al. 2015).

The 5′-terminal stem-loop structures function in a virus-specific manner. The replacement of this stem-loop in the SINV genome (but not the upstream-located nucleotides) by that derived from SFV makes the chimeric virus nonviable, but such RNA is still capable of functioning as a template for negative-strand RNA synthesis in a cell-free in vitro replication system (Frolov et al. 2001). The replacement of both the SINV-specific stem-loop and the unpaired nucleotides located upstream by the corresponding 5′-terminal sequence of SFV genome abolishes both viral replication and RNA synthesis in vitro (Gorchakov et al. 2004). Such replacement leads to rapid viral evolution, which results in the generation of new AU-rich repeating sequences upstream of the heterologous stem (Gorchakov et al. 2004), which are not involved in base pairing. Similarly, modifications of the 5′-terminal stem in VEEV G RNA, which either increase or decrease its free energy, have strong negative effects on viral replication and induce evolution of the 5′UTR that results in the acquisition of additional AU-rich sequences (Kulasegaran-Shylini et al. 2009a).

The 5′UTRs of alphavirus G RNAs (i) determine translation of viral nonstructural polyproteins and (ii) function as a part of the promoter for the negative-strand RNA synthesis. (iii) The complement of the negative-strand 5′ end also acts as a key element in the promoter for positive-strand RNA synthesis. Thus, 5′UTRs exhibit multiple activities in viral replication, while keeping the Cap(0) inefficiently recognized by cellular IFIT1. In each of these processes, the 5′UTR likely functions not very inefficiently and can be further optimized for each particular function. However, the better performance of 5′UTR in one process may alter its efficiency in others. For example, SINV-specific defective interfering (DI) RNAs replicate more efficiently after acquiring a $tRNA^{Asp}$ sequence at their 5′UTR (Monroe and Schlesinger 1983), but the same tRNA-derived sequence rapidly evolves in the context of the viral genome to mediate other functions. Another example is that natural isolates of SINV and VEEV have evolved during passage in cultured cells. Mutations in their 5′-terminal stem-loop sequences have partially destabilized stems and led to a more efficiently replicating phenotype (Reynaud et al. 2015; Kulasegaran-Shylini et al. 2009b). On the other hand, these mutations increased viral sensitivity to IFIT1 and attenuated them in vivo (Reynaud et al. 2015; Hyde et al. 2014).

The 5′UTRs of the SG RNAs in all of the alphaviruses are also predicted to contain relatively small stem-loop structures, and they all have AU at the start, as do the G RNAs. This folding appears to play the same important role in shielding Cap (0) at the 5′ termini. Moreover, SINV variants having the 5′UTR in G RNA replaced by that derived from SG RNA were shown to function efficiently in G RNA replication, but only in vertebrate, and not in insect, cells (Fayzulin and Frolov 2004). This cell-specific RNA replication suggests the involvement of host factors in function of the 5′UTR as a promoter during alphavirus RNA replication.

The accumulated data suggest that the activity of the 5′-terminal core promoter in alphavirus G RNA depends on the presence of another RNA element, which was termed the replication enhancer. This sequence element (51-nt CSE) is relatively conserved among different alphavirus representatives and is located in the nsP1 coding sequence ~150-nt downstream of the Cap. The 51-nt CSEs of all of the alphaviruses are predicted to form two closely located stem-loop structures (Strauss and Strauss 1994). Clustered mutations in SINV 51-nt CSE, which did not change the encoded amino acid sequence of nsP1, decreased rates of viral replication by 4–6 orders of magnitude (Fayzulin and Frolov 2004). Similar extensive mutagenesis of the 51-nt CSE in VEEV genome made the virus essentially nonviable (Michel et al. 2007). However, the absence of CSE can be at least partially compensated by the accumulation of mutations in nsP3 and nsP2 proteins, which make viruses capable of more efficient replication (Fayzulin and Frolov 2004; Michel et al. 2007), albeit not at the wt levels. By now, the mechanism of interaction between the core promoter encoded by the 5′UTR and 51-nt CSE is not well understood. However, indirect experimental data suggest that in the alphavirus dsRNA replication intermediate, the 3′ end of the negative strand does not form a duplex with the 5′ terminus of the positive strand. Thus, in RNA replication, the ends of both strands appear to function as ssRNAs and are likely folded into secondary structures (Kulasegaran-Shylini et al. 2009a). If this is indeed the case, the RNA secondary structure brings the 5′ stem-loop and the 51-nt CSE into close proximity (Gorchakov et al. 2004).

Packaging signal (PS) is a *cis*-acting RNA element in alphavirus G RNA that determines its selective packaging into released virions in the presence of high concentrations of cellular mRNAs and viral SG RNAs in particular. PS structure and functions were intensively investigated for VEEV and SINV (Kim et al. 2011b; Frolova et al. 1997; Weiss et al. 1994, 1989). In these genomes, PSes are located in the nsP1 coding sequence. They are represented by combinations of stem-loop structures, which have additive functions in G RNA packaging. The key elements of these stem-loops are GGG sequences, which are located at the same position in the base of the loops. Destabilization of the PS-specific stem-loops by point mutations or substitutions of the nucleotides in the GGG sequences result in the release of viral particles containing mostly SG RNAs instead of viral genomes (Kim et al. 2011b). However, members of SFV complex, which includes CHIKV, do not contain PS in the nsP1 gene, but instead, the PS is located between nt 2500 and 3100 in the nsP2 coding sequence (Kim et al. 2011b). Published analyses of the SFV DI RNA sequences and detailed investigation of the Ross River virus (RRV)-specific PS suggest that the main PS element in the G RNAs of these viruses and CHIKV lies between nt 2761 and 2994 (White et al. 1998; Frolova et al. 1997; Lehtovaara et al. 1981). While PSes determine the specificity of alphavirus G RNA packaging, they are probably not the only determinants in this process. Chimeric alphaviruses, which combine CHIKV structural protein genes with nonstructural genes derived from heterologous alphavirus species are viable, can replicate both in vivo and in vitro to relatively high titers, but demonstrate highly attenuated phenotype (Kim et al. 2011a; Wang et al. 2008, 2011).

3 Alphavirus Nonstructural Proteins and Regulation of RNA Synthesis

3.1 Nonstructural Polyprotein Synthesis and Processing

The alphavirus genes encoding nonstructural proteins (nsPs) exhibit higher levels of conservation than their structural genes, because of the defined enzymatic functions of the nsPs in viral RNA synthesis. NsP1, nsP2, nsP3, and nsP4 are translated directly

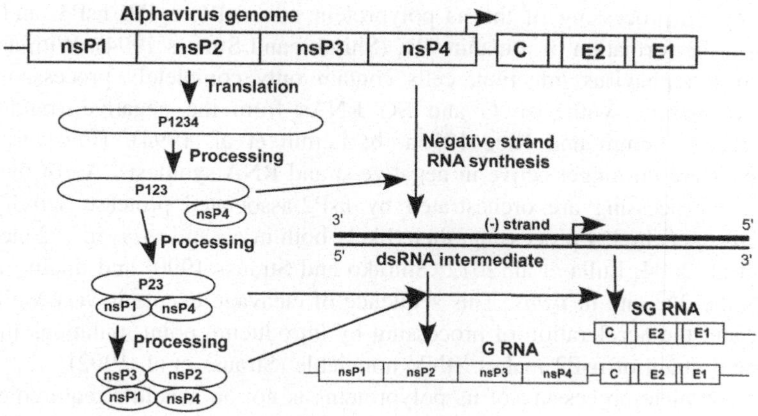

Fig. 2 Schematic presentation of alphavirus ns polyprotein processing and regulation of virus-specific RNA synthesis

from G RNA. Initially, they are synthesized as two ns polyproteins, P123 and P1234 (Fig. 2), which are sequentially processed into individual nsPs by nsP2-associated protease activity (Hardy and Strauss 1989). In many alphavirus species, synthesis of P1234 is determined by the ribosome read through the opal codon located at the carboxy terminus of the nsP3 coding sequence (Strauss et al. 1983). However, natural isolates of alphaviruses do not all have an opal termination codon, and some contain an open reading frame through the entire P1234 (Takkinen 1986). A vast majority of natural CHIKV isolates contain an opal codon between nsP3 and nsP4 (Scholte et al. 2013). However, passaging of CHIKV in vertebrate cells often leads to the replacement of the opal codon by that coding for arginine ((Mounce et al. 2017) and IF unpublished data). The role of the opal codon has only been investigated for the O'nyong'nyong virus (ONNV), a close relative of CHIKV. Its presence increases viral infectivity for mosquitoes. (Myles et al. 2006).

The sequential ns polyprotein processing leads to the formation of partially processed intermediates and regulates the synthesis of viral negative-strand RNA, as well as viral G and SG RNAs. The originally produced P1234 is incapable of any RNA synthesis. After cleavage of the 3^4 cleavage site, the nsP4 release makes P123 + nsP4 replication complex (RC) that is active in synthesizing the negative-strand RNA intermediate (Hardy and Strauss 1989). This intermediate is present in cells as a double-stranded RNA (dsRNA) duplex with viral G RNA. In SFV, the synthesized negative strand was shown to contain a poly(U) sequence, suggesting that the poly(A) tail might be an essential element of the promoter (Sawicki and Gomatos 1976). The P123 + nsP4 complex either remains inactive or very inefficient in positive-strand RNA synthesis (Hardy and Strauss 1989; Shirako and Strauss 1994). However, processing of P123 to nsP1 and P23 changes the RC's template specificity, and nsP1 + P23 + nsP4 complex is able to synthesize positive-strand RNAs (Lemm and Rice 1993a, b; Lemm et al. 1994). Ultimately, the complete processing of the ns polyprotein into nsP1, nsP2, nsP3, and nsP4 results in the formation of a mature RC (Shirako and Strauss 1994). Within a few hours post alphavirus infection, cells contain only completely processed nsPs, which efficiently synthesize G and SG RNAs from the negative-strand RNA intermediate (Lemm and Rice 1993a, b; Lemm et al. 1994). However, these complexes are no longer active in negative-strand RNA synthesis. All of the steps in P1234 processing are orchestrated by nsP2-associated protease activity that functions in *cis* in 3^4 processing, then, likely both in *cis* and *trans* in 1^2 cleavage (Kim et al. 2004; Lulla et al. 2012; Shirako and Strauss 1990) and finally rapidly cleaves the 2^3 site in *trans*. This sequence of cleavage events is very well optimized, and the acceleration of processing by introducing point mutations into the protease domain of nsP2 makes SINV nonviable (Strauss et al. 1992).

The complete processing of ns polyproteins is not an absolute requirement for alphavirus replication. SINV and SFV cleavage mutants, which are incapable of processing either 2^3 or both 1^2 and 2^3 cleavage sites, have been generated (Mai et al. 2009; Kim et al. 2004; Gorchakov et al. 2008a; Lulla et al. 2013). Depending on the viral species used and alterations introduced into cleavage sites, these variants require either only 1–2 adaptive point mutations in nsPs for their viability or no compensatory mutations at all. Defined alterations of 2^3 cleavage made G RNA synthesis even more efficient than in wt alphavirus infections (Kim et al. 2004; Gorchakov et al. 2008a). However, these mutations in 2^3 and 1^2^3 cleavage sites made the OW alphaviruses incapable of producing free, completely processed nsP2. Consequently, the designed P123 cleavage mutants did not exhibit cellular transcription inhibitory functions and became very potent type I IFN inducers (Gorchakov et al. 2008a; Lulla et al. 2013). This in turn affected their ability to develop spreading infection and a cytopathic effect in vertebrate cells, which have no defects in either type I IFN production or signaling (Gorchakov et al. 2008a). Alterations of ns polyprotein cleavage also decreased SINV and SFV replication rates in mosquito cells by a few orders of magnitude (Gorchakov et al. 2008a; Kim et al. 2004). In case of CHIKV, the effects of alterations in ns polyprotein processing were not investigated. However, strong similarities in the CHIKV replication process and virus–host interactions of other Old World alphaviruses suggest that mutations in the cleavage sites may present a possible means of CHIKV attenuation. This may lead to the development of mutants incapable of interfering with the host innate immune response.

3.2 Functions of Alphavirus Nonstructural Proteins in RNA Synthesis and Virus–Host Interactions

Alphavirus genomes encode only four nonstructural proteins (Fig. 3). Most of them have multiple functions in the synthesis of viral G and SG RNAs and in the modification of the intracellular environment to promote efficient viral replication.

3.2.1 Nonstructural Protein 1, nsP1

The known functions of nsP1 include binding of RC to the inner surface of plasma membrane (Laakkonen et al. 1998) and capping of viral G and SG RNAs. The alphavirus-specific mechanism of RNA capping is unique and determined by methyltransferase (MTase) and guanylyltransferase (GTase) enzymatic activities of nsP1 (Ahola and Kaariainen 1995; Ahola et al. 1997). This protein mediates a cascade of enzymatic reactions that require GTP and S-adenosyl methionine as a donor to synthesize m^7GMP. The latter intermediate remains covalently linked to nsP1, forming m^7GMP-nsP1, and then, m^7GMP is translocated to viral positive-strand RNAs by 5′-5′ linkage. G and SG RNAs are pre-activated for this reaction by the RNA triphosphatase activity of nsP2 (Vasiljeva et al. 2000) that leaves the PP group at their 5′ termini. The efficiency of nsP1-/nsP2-mediated capping is far from being absolute, and noncapped viral genomes are abundantly present in alphavirus-infected cells and released viral particles (Sokoloski et al. 2015). Most of these RNAs contain the 5′-terminal monophosphate (Sokoloski et al. 2015). So far, targeting of RCs to the plasma membrane and capping activities are the most well studied, but may not be the only nsP1-specific functions. For example, this protein was also proposed to induce rearrangements of the plasma membrane and interfere with tetherin-mediated inhibition of CHIKV particle release (Jones et al. 2013).

3.2.2 Nonstructural Protein 2, nsP2

NsP2 is the largest alphavirus nonstructural protein, and to date, its structural domains have not been completely defined or characterized in terms of their functions (Fig. 3). The list of known nsP2 activities in viral replication is continuously growing.

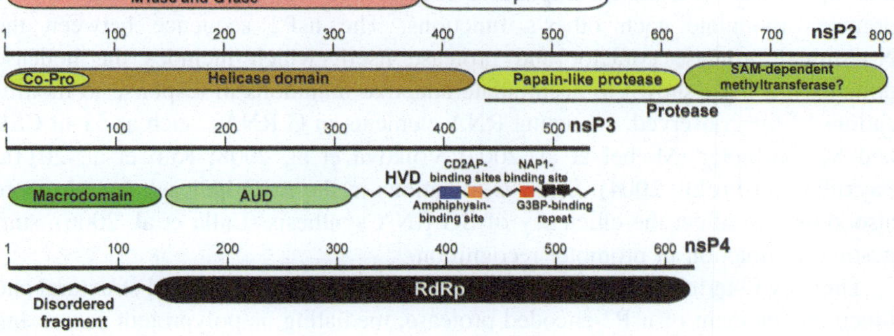

Fig. 3 Schematic presentation of functional domains and binding motifs in CHIKV nsPs

Similar to other alphaviruses, CHIKV-specific nsP2 functions as a papain-like cysteine protease in the processing of nonstructural polyprotein precursors P123 and P1234 (Rausalu et al. 2016). Protease function is associated with the C-terminal fragment of this protein, which contains two large domains: the catalytic protease domain and the very C-terminal S-adenosyl-L-methionine (SAM)-dependent RNA methyltransferase-like (SAM MTase-like) domain (Russo et al. 2006; Strauss et al. 1992; Shin et al. 2012) (see also PDB ID: 3TRK for CHIKV protease). However, the proteolytic activities of at least SFV- and SINV-specific nsP2 proteases are also determined by the very N-terminal ~100-aa-long fragment of nsP2 and the macrodomain of nsP3. The N-terminal fragment of nsP2 was proposed to function as a protease cofactor, similar to HCV 2A or flavivirus NS2B proteins (Vasiljeva et al. 2003; Atasheva et al. 2007; Lulla et al. 2012). Consequently, mutations and large insertions into the N-terminal peptide had strong negative effects on the rates of P123 and P23 processing. Thus, it is reasonable to expect that the function of the CHIKV protease also depends on integrity of the N-terminus of nsP2. The macrodomain of nsP3 is in turn a critical determinant of 2^3 cleavage, and this makes the processing of this site dependent on macromolecular assembly in addition to being determined by the specific amino acid sequence of the 2^3 cleavage site (Lulla et al. 2012).

The N-terminal domain (~430 aa) of alphavirus nsP2 encodes helicase, NTPase, and RTPase functions (Rikkonen et al. 1994; Karpe et al. 2011; Das et al. 2014; Vasiljeva et al. 2000). The RNA triphosphatase activity prepares G and SG RNAs for capping reactions, mediated by nsP1, by creating a di-phosphate group at the 5′ terminus (Vasiljeva et al. 2000). The helicase component of CHIKV nsP2 belongs to the SF1 superfamily (Gorbalenya et al. 1989). It contains two RecA-like domains, two classical Walker motifs (Walker A and Walker B), and an arginine finger, which represent characteristic functional elements of SF1 helicases. NsP2 proteins of SFV and CHIKV exhibit NTPase and dNTPase, helicase and RNA-annealing activities (Gomez de Cedron et al. 1999; Das et al. 2014; Vasiljeva et al. 2000), which play indispensable roles in viral RNA replication. Compared to the entire nsP2 protein, the N-terminal, helicase-encoding domain alone demonstrates very low enzymatic activity (Das et al. 2014). This is an additional indication that alphavirus nsP2 has a complicated structure, in that the protease and helicase domains modulate each other's functions. The nsP2 sequence between the N-terminal protease cofactor and protease itself, which includes the helicase domain, has been shown to accumulate adaptive mutations in response to modifications of the conserved, *cis*-acting RNA elements in G RNAs, such as 51-nt CSE and SG promoter (Michel et al. 2007; Volkova et al. 2008; Kim et al. 2011a; Fayzulin and Frolov 2004). Some *ts* mutations in the nsP2 helicase domain were also shown to affect the efficiency of SG RNA synthesis (Lulla et al. 2006), suggesting its function in promoter recognition.

The very C-terminal SAM MTase-like domain of nsP2 is a critical structural and functional element of nsP2-encoded protease, mediating ns polyprotein processing and regulating the synthesis of virus-specific RNAs during the replication cycle (Shirako and Strauss 1994; Lemm and Rice 1993b). However, this domain also plays a critical role in nsP2-specific nuclear function(s), a distinguishing characteristic of the Old World alphaviruses studied to date. Point mutations in the SAM MTase-like domain make CHIKV and other Old World alphaviruses dramatically less cytopathic (Dryga et al. 1997; Frolov et al. 1999; Perri et al. 2000; Utt et al. 2015). The negative effect of these mutations on viral cytopathogenicity likely has two components: (i) The defined mutations in the SAM MTase-like domain make Old World alphaviruses and their replicons (defective viral genomes, which do not express viral structural proteins) incapable of inducing transcriptional shutoff in infected cells, and (ii) most of the identified mutations also strongly downregulate synthesis of virus-specific RNAs. While the crystal structures of VEEV and SINV nsP2 protease domains have been resolved (Shin et al. 2012; Russo et al. 2006), the mechanisms of the mutation effects on RNA replication and nsP2-mediated cytopathogenicity remain unclear. This is due to the fact that Old World alphavirus nsP2 is present in both nuclear and cytoplasmic compartments, where it interacts with a wide range of cellular and viral proteins in assembly of viral RCs and other complexes (Atasheva et al. 2007). Thus, dissection of the mutation effects on particular interactions and processes in viral and RNA replication is complicated.

In vertebrate cells infected with Old World alphaviruses, such as CHIKV, SINV, or SFV, a large fraction of the nsP2 protein is detected in the nuclei (Frolova et al. 2002; Rikkonen et al. 1992; Peranen et al. 1990). Moreover, if expressed alone, out of context of the P123 polyprotein, almost the entire pool of nsP2 is translocated into the nucleus (Frolov et al. 2009). In SFV-infected cells, nuclear and cytoplasmic forms of nsP2 were found to interact with different nsP2-specific mAbs and, thus, were proposed to have distinct conformations (Kujala et al. 1997). However, this can also be explained by different accessibility of antigenic sites in distinct nsP2 complexes. NsP2 proteins of the Old World alphavirus species contain either one or two putative nuclear localization signals (NLS). Mutations in SFV nsP2-specific NLSes affect translocation of these proteins into the nucleus (Fros et al. 2013). However, mutations in both putative NLSes of SINV nsP2 had no effect on its ability to accumulate in the nuclei, suggesting that this nsP2 protein is likely capable of using other means of nuclear transport (Frolov et al. 2009; Mayuri et al. 2008). Mutations introduced in putative CHIKV nsP2-specific NLS, which has no classical amino acid sequence, had a profound negative effect on RNA replication (Fros et al. 2013). This indicates that besides affecting the putative NLS, these mutations modified the overall conformation of either SAM MTase-like domain or even the entire nsP2 protein.

Expression of nsP2 proteins derived from CHIKV, SFV, and SINV out of context of the ns polyprotein is highly cytotoxic, because these proteins efficiently induce global transcriptional shutoff and possibly other changes in cell physiology (Akhrymuk et al. 2012). Within 6 h post-infection with CHIKV and other Old World alphaviruses, transcription of cellular messenger and rRNAs is inhibited to undetectable levels (Akhrymuk et al. 2012; Garmashova et al. 2006, 2007). Extensive random insertion mutagenesis studies on SINV nsP2 demonstrated that the mutations, which make it incapable of causing transcription inhibition, and consequently, cell death, have a clustered distribution in the helicase, protease, and

SAM MTase domains (Frolov et al. 2009). This supports the hypothesis that all of the nsP2 domains not only cooperatively function in viral RNA replication, but also determine nsP2-specific activities in other processes of virus–host interactions, such as modifications of nuclear function(s). Some of the mutations in SINV nsP2 did not affect putative NLSes, but made the protein incapable of accumulating in nuclei. Accordingly, these mutants became incapable of inhibiting cellular transcription and lost their cytopathic phenotype (Frolov et al. 2009). Other mutations, which were closely clustered in the SAM MTase-like domain, had no effect on the transport of nsP2 to the nucleus, but made it incapable of inducing transcriptional shutoff. These data suggested that interaction with nuclear factors and subsequent transcription inhibition are the critical functions of the SAM MTase-like domain (Frolov et al. 2009). Similar results were noted in studies where mutated SINV and SFV replicons and viruses capable of persistent replication in cells without inducing profound cytopathic effect were selected (Frolov et al. 1999; Dryga et al. 1997; Perri et al. 2000; Frolov and Schlesinger 1994; Liljeström and Garoff 1991). In these variants, the attenuating point mutations were also identified in the SAM MTase-like domain. In contrast to other alphaviruses, selection of noncytopathic CHIKV replicons having attenuating mutations in nsP2 or other nsPs was found to be more challenging. Multiple mutations in the SAM MTase-like domain were required for producing a noncytopathic phenotype (Utt et al. 2015).

The most recent studies on CHIKV and SINV led to the identification of a short, highly variable peptide (VLoop) located on the surface of the SAM MTase-like domain of nsP2 that plays a critical role in the transcription inhibitory function of the protein (Akhrymuk et al. 2018a, b). Clustered mutations introduced into this peptide had no negative effect on RNA and viral replication, but these mutants were no longer capable of inducing transcriptional shutoff. This made them very potent type I IFN inducers. Consequently, they were cleared without developing cytopathic effect from IFN-competent rodent cells or could develop persistent replication in the cells that were defective in type I IFN induction or signaling. Taken together, the data generated on a variety of the Old World alphaviruses, including CHIKV, demonstrated that nuclear functions of nsP2 can be altered by either the mutations in the SAM MTase-like domain or by alterations in the P2^3 cleavage site (Gorchakov et al. 2008a), which make nsP2 incapable of translocation to the nucleus.

After finding that SINV, SFV, and CHIKV infections induce rapid degradation of RPB1, the catalytic subunit of cellular DNA-dependent RNA polymerase, it was tempting to speculate that nsP2-induced transcriptional shutoff was the result of the nsP2-associated protease activity. However, this was not the case. The Old World alphaviruses exploit a different pathway, where the nuclear form of nsP2 induces polyubiquitination of the RPB1 subunit in the context of an elongating RNAPII complex and ultimately directs it into proteasome degradation (Akhrymuk et al. 2012). This makes alphavirus-specific mechanism of RPB1 degradation similar to that previously described for the transcription-coupled repair (TCR) pathway (Svejstrup 2003). Accordingly, nsP2 with its protease activity inactivated by a point mutation in the catalytic center induced RPB1 degradation as efficiently as its wt counterpart. However, mutations in the helicase domain or in the C-terminal SAM MTase-like domain, which were previously identified in noncytopathic replicons, inactivated nsP2 function in RPB1 degradation (Akhrymuk et al. 2012; Garmashova et al. 2006). Since alphavirus helicases demonstrate high affinity to dsDNA (EF unpublished data), the current model of transcriptional shutoff suggests that nsP2 binds to cellular DNA and stalls elongating RNAPII. The SAM MTase-like domain may, in turn, mediate interaction with cellular factors, which are involved in the RPB1 polyubiquitination process. However, this hypothesis needs an additional experimental support.

Other important points about the nuclear function of nsP2 are as follows.

(1) NsP2 proteins of CHIKV and other Old World alphaviruses do not induce degradation of RPB1 in mosquito cells (Akhrymuk et al. 2012). This provides a plausible explanation for persistent, noncytopathic replication of alphaviruses in insect cells.
(2) Upon infection of vertebrate cells with the Old World alphaviruses, RPB1 degradation is completed by 6 h post-infection, before modifications of cellular biology, such as apoptosis, autophagy, and inhibition of STAT1 phosphorylation, develop (Akhrymuk et al. 2012; Frolov et al. 2012). Thus, RPB1 degradation and transcription inhibition appear to be the critical mechanisms applied by the Old World alphaviruses for evasion of the innate immune response.
(3) In alphavirus-infected cells, inhibition of transcription results in cytoplasmic retention of numerous nuclear trafficking factors, such as hnRNPs, DHX9, NCL, and others, which normally shuttle between the cytoplasm and the nucleus (Akhrymuk et al. 2012). This cytoplasmic accumulation of nuclear proteins is not a specific function of viral nsPs and is similar to the effects induced by nonspecific transcription inhibition caused by ActD (Siomi et al. 1997).
(4) CHIKV nsP2 was also proposed to inhibit IFN-stimulated JAK-STAT signaling by interfering with STAT1 phosphorylation and translocation into the nucleus (Fros et al. 2010, 2013). This may be an additional mechanism of nsP2-mediated inhibition of the innate immune response. However, in SINV-infected cells, this function directly depends on the nsP2-induced inhibition of cellular transcription and can be detected only at late times post-infection, when cell already demonstrates severe alterations of nuclear functions and strong signs of cytopathic effect (Frolov et al. 2012).

3.2.3 Nonstructural Protein 3, nsP3

In contrast to other viral nonstructural proteins, nsP3 still has no assigned enzymatic functions in alphavirus RNA replication. NsP3 has three structural domains, of which the N-terminal macrodomain (Malet et al. 2009) and the following zinc-binding (Shin et al. 2012), alphavirus unique domain (AUD) are conserved among all of the alphaviruses (Fig. 3). The macrodomain, also referred to as the X-domain, is homologous to similar domains found in nonstructural proteins of many other positive-strand RNA viruses, and in some bacterial and cellular proteins (Rack et al. 2016). Based on available crystal structures of the alphavirus macrodomains and their functional analysis, this domain can bind ADP-ribose, poly (ADP-ribose) and RNA and exhibit low levels of adenosine di-phosphoribose 1″-phosphate phosphatase activity (Malet et al. 2009; Egloff et al. 2006). CHIKV nsP3 macrodomain is able to hydrolyze ADP-ribose groups from mono(ADP-ribosyl) ated proteins (McPherson et al. 2017; Abraham et al. 2018; Eckei et al. 2017). Based on structural and biochemical data, several amino acids were identified as important for binding of ADP-ribose or the hydrolase activity of nsP3 (N24, G32, V33, G112, and Y117) (McPherson et al. 2017; Eckei et al. 2017). CHIKV variants having point mutations in the macrodomain, which reduced its enzymatic activity, demonstrated slower replication rates in mammalian neuronal cells and were attenuated in 2-day-old mice (McPherson et al. 2017). However, there was no clear correlation between the effects of these mutations on enzymatic activity and viral replication. Interestingly, SINV and VEEV variants having mutations in the same fragment (aa 31) of the macrodomain have previously been described (Park and Griffin 2009; Foy et al. 2013b; Michel et al. 2007). The substitutions in this position reverted the negative effects of other mutations introduced either into the very *N*-terminal fragment of SINV nsP3 (Park and Griffin 2009) or into the C-terminal hypervariable domain of VEEV nsP3 (Foy et al. 2013b) and the VEEV 51-nt CSE (Michel et al. 2007). Thus, multiple explanations for effects of macrodomain-specific mutations on alphavirus replication should probably be considered. Moreover, the mutations of N24 in SINV nsP3, which was shown to inhibit mono-ADP-ribosylhydrolase or adenosine di-phosphoribose 1" -phosphate phosphatase activities of CHIKV macrodomain (Malet et al. 2009; Eckei et al. 2017), had no negative effect on viral replication rates, but made these variants inefficient in the inhibition of cellular translation and capable of persistent replication in vertebrate cells (Akhrymuk et al. 2018a). Another known function of the macrodomain is to position the P2^3 cleavage site for access by nsP2 protease during the last step of ns polyprotein processing (Lulla et al. 2012).

Functions of the Zn-binding AUD remain to be understood. The SINV AUD crystal structure has been already resolved (Shin et al. 2012). However, this did not lead to rapid progress in the understanding of its activity in viral replication. It was proposed to have an RNA-binding surface, but the hypothesis about nsP3–RNA interaction needs experimental support. In VEEV and CHIKV, AUDs were shown to accumulate compensatory mutations in response to extensive modifications in the hypervariable C-terminal domain of nsP3, which have deleterious effects on viral replication (Thaa et al. 2015; Foy et al. 2013b; Meshram et al. 2018). Positive effects of such second site mutations are usually specific to BHK-21 cells and do not increase viral replication in most other cell lines of both vertebrate and invertebrate origins (Foy et al. 2013b; Meshram et al. 2018).

The C-terminal nsP3 domain has different lengths and demonstrates exceptionally low levels of conservation between alphavirus species. Therefore, it is termed the hypervariable domain (HVD). It is intrinsically disordered and serves as a hub for binding distinct, virus-specific sets of cellular proteins, which are required for RC assembly and function (Frolov et al. 2017; Kim et al. 2016; Meshram et al. 2018). In different alphaviruses, including CHIKV, HVDs were shown to tolerate insertions of large protein sequences, such as the entire GFP, Cherry and luciferase, without deleterious effects on viral replication (Sun et al. 2014; Gorchakov et al. 2008b; Cristea et al. 2006; Tamberg et al. 2007; Foy et al. 2013a; Frolova et al. 2006). The extended deletions introduced into HVDs of some alphaviruses caused only relatively small, cell-specific negative effects on their replication (LaStarza et al. 1994) and pathogenicity (Vihinen et al. 2001; Hallengard et al. 2014).

In infected cells, two types of nsP3-containing complexes have been identified (Gorchakov et al. 2008b; Kim et al. 2016). One of them, the small, membrane-bound complexes, is associated with plasma and endosomal membranes, and these complexes frequently overlap with dsRNAs, which represent RNA replication intermediates. This overlap suggests a function for these types of complexes in RNA replication. Other cytoplasmic complexes are larger, demonstrate no association with dsRNA, are tightly bound to the cytoskeleton, and have no membrane component. A common characteristic of both types of nsP3 complexes formed by the Old World alphaviruses, including CHIKV, is the presence of high levels of the Ras-GAP SH3 domain-binding proteins, G3BP1 and G3BP2. A few hours post-infection, essentially the entire cellular pools of both proteins relocalize into nsP3-specific complexes (Gorchakov et al. 2008b). G3BP–nsP3 interaction is mediated by the N-terminal NTF2-like domain of G3BP and short peptides in the C-terminus of HVD (Fros et al. 2012; Panas et al. 2014; Kristensen 2015). In different Old World alphavirus species, these G3BP-binding peptides demonstrate high levels of conservation. In mosquito cells, the insect-specific homolog of G3BP, Rasputin, interacts with nsP3 HVD, suggesting that this interaction plays a critical role for viral replication in both vertebrate and invertebrate cells (Fros et al. 2015; Gorchakov et al. 2008b; Meshram et al. 2018). Moreover, identification of the same G3BP-interacting peptides in the nsP3 HVD of the recently discovered Eilat alphavirus, which replicates exclusively in insect cells, suggested that this interaction appears to be very evolutionally conserved (Nasar et al. 2012; Meshram et al. 2018). It is also well established that proline-rich motifs that are present in HVDs of the Old World alphaviruses interact with amphiphysins (Neuvonen et al. 2011). Alterations of HVD interaction with amphiphysin 2 (BIN1) caused small, but detectable negative effects on SINV and SFV RNA replication in vitro and on SFV pathogenicity (Neuvonen and Ahola 2009). Several other SINV and CHIKV HVD-binding host proteins have been identified (Frolov et al. 2017; Kim et al. 2016). The CHIKV HVD-specific factors include NAP1L1, NAP1L4, MYBBP1A, BIN1, CD2AP, SH3KBP1, FHL1, and some others (Meshram et al. 2018; Frolov et al. 2017; Mutso et al. 2018). Interestingly, the spectra of host proteins interacting with CHIKV HVD depend on the cell lines used. To date, functions of many HVD-interacting host proteins in CHIKV replication remain to be identified. Importantly, host factors that interact with nsP3 HVDs of the Old World alphaviruses are different from those binding to HVDs of the New

World alphaviruses (Frolov et al. 2017; Kim et al. 2016). Moreover, some of the host proteins bind only to specific viral species. For example, WDR48 was detected only in SINV nsP3 HVD complexes, while NAPL1 and NAPL4 proteins interacted only with CHIKV nsP3 HVD in vertebrate cells. This suggests that even closely related alphaviruses may evolve different mechanisms of usurping cellular environments for their efficient replication.

The described interaction of G3BP with alphavirus HVD was proposed as the virus-specific means of inhibiting the formation of cellular stress granules, which can develop in response to chemical stress or inhibition of translation (McInerney et al. 2005; Panas et al. 2012). However, more recent studies demonstrated that this interaction plays a critical role in the formation of active viral RCs. The proviral role of this interaction is particularly evident for CHIKV (Scholte et al. 2015; Kim et al. 2016). In the absence of its interaction with G3BPs, which was achieved by either knockout of both G3BP1 and G3BP2 expression or by deletion of the G3BP-specific repeat in CHIKV nsP3 HVD, viral replication was reduced to almost undetectable levels (Kim et al. 2016). This makes CHIKV somewhat unique, because similar alterations of G3BP–HVD interaction in other Old World alphaviruses affected their replication rates by 2–3 orders of magnitude, but did not make them nonviable. Thus, other HVD-interacting host factors are capable of supporting some low levels of other alphavirus replication, but not of CHIKV. However, the inability of CHIKV to replicate in the absence of G3BP–HVD binding certainly does not mean that other HVD-binding host factors are not essential. The extensive mutagenesis of CHIKV HVD without alteration of G3BP-binding motifs or replacement of CHIKV HVD by HVDs of other alphaviruses made virus nonviable in both vertebrate and mosquito cells (Meshram et al. 2018), indicating that other binding host factors are as important for viral replication as G3BPs. Among them, the SH3 domain-containing host proteins appear to play more proviral role in insect cells, and NAP1 family members are likely more important for CHIKV replication in the cells of vertebrate origin (Meshram et al. 2018).

3.2.4 Nonstructural Protein 4, nsP4

Compared to other nsPs, nsP4 is present in alphavirus-infected cells at lower concentrations. For most of the alphavirus species, nsP4 is translated through the opal codon located between nsP3 and nsP4 coding sequences, and unless being in complex with other nsPs, the free form of nsP4 also undergoes degradation by the N-end rule pathway (de Groot et al. 1991). In the context of initially synthesized P1234 polyprotein, nsP4 does not exhibit any enzymatic activities. However, upon the release by nsP2-mediated cleavage, it functions as an RNA-dependent RNA polymerase (RdRp), synthesizing the negative-strand RNA intermediate and later, both G and SG RNAs. To date, nsP4 of any alphavirus has not been produced at the levels sufficient for structural studies, and the 3D structure remains to be determined. However, this protein is predicted to have two distinct domains. The N-terminal domain is specific for alphaviruses. This ~140 amino acid-long fragment demonstrates lower levels of conservation among alphaviruses than the rest of nsP4. It also appears to be intrinsically disordered, and point mutations introduced into this fragment strongly affect SINV replication, leading to the selection of compensatory mutations in other nsPs (Rupp et al. 2011). Thus, the N-terminal domain may determine RC assembly and function by mediating interactions of nsP4 with other viral and host components.

Based on computer modeling, the second ~500-aa-long nsP4-specific domain has a structure that is similar to RdRps of other RNA$^+$ viruses (Tomar et al. 2006; Rubach et al. 2009). So far, neither the purified full-length nsP4 nor its truncated versions with deletions in the N-terminal domain have exhibited de novo RNA synthesis unless the in vitro reaction was supplemented with a cellular membrane fraction that contained the P123 polyprotein (Rubach et al. 2009). Both the C-terminal nsP4 domain and complete SINV nsP4 were shown to have adenylyltransferase activity in vitro (Tomar et al. 2006; Rubach et al. 2009). This activity may play a critical role in saving the poly(A) tails of alphavirus G and SG RNAs at the length required for RNA stability and function of the 3′ CSE as a promoter for negative-strand RNA synthesis.

4 Replication Complexes

4.1 Compartments of Alphavirus RNA Replication

Staining of alphavirus-infected cells with dsRNA-specific Abs readily detects formation of dsRNAs, which serve as replication intermediates in the synthesis of viral G and SG RNAs (Gorchakov et al. 2008b; Kim et al. 2016). The dsRNAs can be also isolated from SINV-infected cells using biochemical techniques (Mai et al. 2009). Formation of dsRNAs is an essential step in alphavirus replication. On the other hand, dsRNA is also one of the pathogen-associated molecular patterns (PAMPs) that is produced during replication of many viruses and potentially sensed by cellular pattern recognition receptors (PRRs). Thus, accumulation of alphavirus dsRNAs can induce the pathways that trigger the antiviral response in the already infected cells and cell signaling. Such signaling can activate the antiviral state in yet uninfected cells and make them not susceptible to subsequent rounds of viral infections. The hallmark of the induced cell signaling is the release of type I IFN that activates more than two hundred interferon-stimulated genes (ISGs), whose products can interfere with different steps of the alphavirus replication cycle. Thus, the spread of alphavirus infections and development of high titer viremia, which is essential for viral transmission to mosquito vectors, are dependent not only on efficient viral replication but also on downregulation of the innate immune response. Similarly, the appearance of unprotected dsRNA in the cytoplasm of insect cells could result in efficient formation of virus-specific siRNAs, which interfere with alphavirus replication (Blair 2011). One of the means of minimizing induction of the antiviral response is making dsRNAs not accessible to cellular PRRs, such as RIG-I, MDA5, and PKR. This is achieved by isolating the virus-specific dsRNAs into membrane-bound structures, which, in case of alphavirus infections, are termed spherules. These membrane invaginations are formed by all of the studied alphaviruses and have a similar size of ~50–70 nm (Frolova et al. 2010; Spuul et al. 2010). Expression of viral nonstructural proteins from the nonviral vectors does not lead to spherule formation. Moreover, recent studies have demonstrated that the spherule size is determined by the length of the template RNA supplied *in trans* for dsRNA synthesis (Kallio et al. 2013). Thus, viral RNA replication and formation of dsRNA intermediates are prerequisites of spherule formation, and they are equally efficiently produced by wt SINV and its 2^3 and 1^2^3 cleavage mutants, indicating that P123 and nsP4 determine their generation (Frolova et al. 2010). Since spherule discovery a few decades ago (Froshauer et al. 1988; Friedman et al. 1972; Grimley et al. 1968, 1972), endosomes and early lysosomes were considered to be the sites of alphavirus-specific RNA synthesis. However, this is not exactly the case. At early times post-alphavirus infection, both spherules and dsRNAs accumulate at the plasma membrane (Frolova et al. 2010; Spuul et al. 2010; Kim et al. 2016; Kujala et al. 2001). This compartmentalization correlates with the efficient targeting of nsP1, the N-terminal component of the ns polyprotein, to the plasma membrane. As infection progresses, endocytosis mediates relocalization of a significant fraction of spherules to the endosomal and lysosomal membranes (Froshauer et al. 1988).

Thus, initially, spherules are located on the external surface of the plasma membrane and are connected to the cytoplasm via a very narrow neck (Froshauer et al. 1988). The dsRNAs are packed inside the spherule cavities (Gorchakov et al. 2008b), and after permeabilization of the membrane, specific Abs readily detect dsRNAs at the cell surface (Frolova et al. 2010). To date, the presence of nsP1, nsP2, and nsP3 in the spherules was experimentally confirmed (Frolova et al. 2010). However, their detection required signal amplification, suggesting that each spherule likely contains only one RC. On the other hand, alphavirus nsPs are abundantly present on the cytoplasmic side of the necks (Froshauer et al. 1988). These nsPs and host factors involved in RNA replication form dense structures that are readily detectable by EM after negative staining or staining with gold-labeled Abs (Froshauer et al. 1988; Frolova et al. 2010). Such plugs likely make the dsRNA sensing by PRRs even more challenging. At later times post-infection, some spherules are internalized by the cells and remain on the internal surface of modified endosomes and lysosomes. Traditionally, these organelles are termed cytopathic vacuoles (CPV-I). Thus, at any time post-infection, dsRNAs are isolated in the membrane compartments. However, RIG-I, MDA5, and PKR are still capable of detecting dsRNA during alphavirus replication, suggesting that likely not all dsRNA molecules are packed. Experimental data demonstrate that CHIKV infection induces PKR phosphorylation (White et al. 2011), which then causes detectable inhibition of cellular translation. Phosphorylation of IRF3 via the MAVS-dependent mechanism in CHIKV-infected cells is also indirect evidence that produced dsRNAs are detected by RIG-I and MDA5 (White et al. 2011).

Experiments with other alphaviruses demonstrated that dsRNA sensing by cellular PRRs is likely to be common for all of the members in this genus (Akhrymuk et al. 2016). However, activation of type I IFN is usually poorly detectable because of the virus-induced transcriptional and translational shutoffs, which nonspecifically inhibit activation of the antiviral response in vitro. Viral mutants that lack nuclear functions, but have no defects in spherule formation, readily produce high levels of type I IFN, and the level of its induction correlates with levels of RIG-I and MDA5 expression (Akhrymuk et al. 2016). Importantly, recent studies have demonstrated that alphavirus nsPs are capable of utilizing cellular mRNA as templates for dsRNA synthesis (Nikonov et al. 2013). Most of mRNAs are shorter than alphavirus genomes and the genomes of defective interfering (DI) RNAs, which are capable of efficient replication. They also lack alphavirus-specific CSEs. Therefore, there is a possibility that produced nonviral dsRNAs induce the formation of membrane spherules less efficiently and some of them remain exposed to PRRs.

4.2 Assembly of Viral Replication Complexes

RC formation is one of the most intriguing processes in alphavirus biology. Despite a considerable progress in understanding the mechanism of their genesis, many critical aspects of RC assembly and function and the mechanism of spherule formation still remain to be understood.

It is clear that viral nonstructural proteins play indispensable roles in RNA replication, and differential processing of the ns polyprotein regulates synthesis of the negative-strand RNA intermediate (in dsRNA form), G and SG RNAs. Structural genes can be deleted from viral G RNAs or replaced by heterologous genes (Liljestrom 1994; Frolov et al. 1996). Such defective genomes, termed replicons, remain capable of self-replication and transcription of the subgenomic RNA. Thus, viral structural proteins are generally dispensable in viral RNA synthesis. However, this does not necessarily mean that structural proteins, and capsid protein in particular, have no effect on the efficiency of RNA synthesis and viral replication. The recent study has identified new RNA elements in SINV G RNA that interact with capsid protein, and mutations in these sequences have negative effects on viral replication and pathogenesis (Sokoloski et al. 2017). In turn, cellular proteins are critical players in viral G RNA replication and transcription of SG RNA. SINV nsP2 and nsP4, and nsP3 of CHIKV, VEEV, and SINV were fused with different tags in the context of infectious virus or replicons without deleterious effects on RNA replication (Atasheva et al. 2007; Frolova et al. 2006; Spuul et al. 2010; Foy et al. 2013a; Cristea et al. 2006, 2010). The affinity-purified complexes isolated from the infected cells contained all of the nsPs and large numbers of cellular factors, which were identified by mass spectrometry (Gorchakov et al. 2008b; Cristea et al. 2006, 2010). These experiments provided important information, but it remained unclear with which particular nsP each host protein interacted. Identification of host factor functions was also complicated, because entire families of cellular proteins, such as G3BP, 14-3-3, FXR, and numerous hnRNPs, were identified. The use of multiple family members suggested high redundancy in host proteins' functions, and thus, knockout or knockdown of individual protein expression often has only minor effects on viral replication, if any at all. It is also possible that host factors function in a cell-specific mode. In the case of Old World alphaviruses, which include CHIKV, the most abundant vertebrate cell-specific proteins co-isolated from infected cells are represented by Ras-GAP SH3 domain-binding proteins G3BP1 and G3BP2 (Frolova et al. 2006; Gorchakov et al. 2008b; Cristea et al. 2006, 2010). SINV-specific complexes isolated from mosquito cells contain high level of Rasputin (Rin), which is a mosquito homolog of G3BPs (Gorchakov et al. 2008b). Rin was also shown to accumulate in complexes formed by CHIKV nsP3 (Fros et al. 2015). G3BPs interact with a short repeating peptide located in the amino terminus of nsP3 HVD (Fros et al. 2012; Panas et al. 2014; Kristensen 2015). CHIKV replication is critically dependent on G3BP–HVD interaction, and the deletion of both G3BP-interacting peptides made virus nonviable (Kim et al. 2016). Wt CHIKV also does not replicate in *G3 bp* dKO cells and replicates less efficiently in cells treated with siRNAs specific to both G3BP1 and G3BP2 (Scholte et al. 2015). Relocalization of the entire pools of G3BP1 and G3BP2 into nsP3-containing protein complexes makes cells unable to form stress granules (Panas et al. 2012, 2014). However, strong antiviral effects of KO of both G3BP genes on replication of all tested Old World alphaviruses, and CHIKV in particular (Kim et al. 2016; Scholte et al. 2015), suggest that inhibition of stress granule formation is likely not the main function of G3BP–nsP3 interaction. One of the possibilities is that G3BPs interact with viral G RNA and clear it from ribosomes in order to recruit it into replication (Scholte et al. 2015). The alternative hypothesis suggests that alphaviruses use their nsP3 HVDs to interact with G3BPs and exploit the natural abilities of these proteins to assemble into higher-order structures for building pre-replication complexes (Fig. 4) (Kim et al. 2016). These preformed complexes recruit viral G RNAs into replication and spherule formation. At early times post-infection, during the RNA amplification stage, G3BP/nsP complexes form readily detectable strips at the plasma membrane (Frolova et al. 2010; Kim et al. 2016). Application of immunostaining and in situ hybridization followed by confocal microscopy allows to distinguish three types of complexes in these strips (Kim et al. 2016). The first type is represented by those containing only G3BPs and unprocessed P123 polyprotein, but not virus-specific RNAs. These are the membrane-bound pre-replication complexes, which are ready for acquiring viral G RNA. The second type of complexes contains P123, G3BPs, and viral ssRNAs, and occasionally, small dsRNA can also be detected. These complexes represent the sites of ongoing synthesis of dsRNA replication intermediates before the beginning of G and SG RNA synthesis. Large dsRNAs were detected by confocal microscopy only in the third type of complexes. The latter complexes are fully functional in viral RNA replication and can actively incorporate BrU into ssRNAs (Frolova et al. 2010). Importantly, signals of immunostaining of dsRNA and nsPs do not exactly co-localize, but often overlap. This is suggestive that nsPs and dsRNAs are closely compartmentalized, but the dsRNA molecules are mostly free of ns proteins.

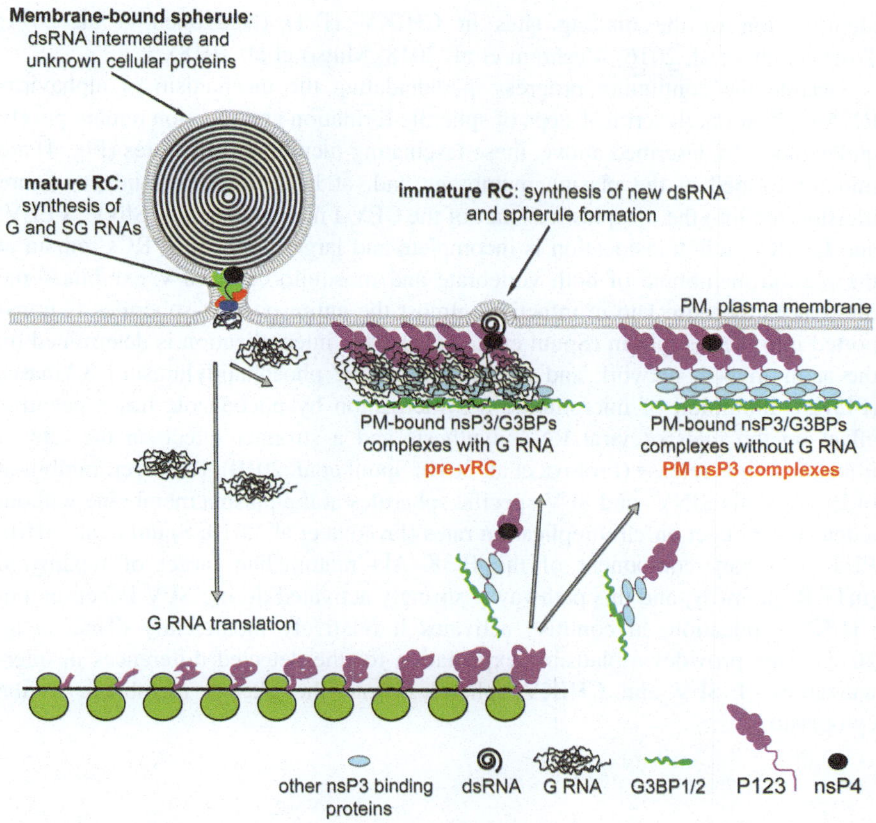

Fig. 4 Schematic presentation of CHIKV spherule and RC formation and roles of viral nsPs and host factors in this process

Co-localization of small amount of nsPs with dsRNA can be conclusively detected only by using signal amplification techniques (Frolova et al. 2010). The presence of other than G3BP host proteins in these complexes has not been extensively investigated yet. The exact functions of G3BPs in CHIKV and other alphavirus replication need to be further investigated, because its understanding will lead to the development of new antiviral screens and means of therapeutic treatment.

Components of the replication complexes and functions of viral and host factors in RNA replication and virus–host interactions are being continuously investigated for a variety of alphaviruses, including CHIKV. HVDs of different members of the alphavirus genus were found to interact with particular sets of RNA-binding proteins, SH3 domain-containing proteins, actin-interacting proteins, BAR domain-containing proteins, and some other host factors (Frolov et al. 2017; Kim et al. 2016). Thus, they were proposed to function as virus-specific hubs in replication complex assembly. Besides G3BPs and Rin, only BIN1, CD2AP, and NAP1 functions were investigated in terms of supporting alphavirus replication and identification of the binding sites in CHIKV HVD (Neuvonen et al. 2011; Tossavainen et al. 2016; Meshram et al. 2018; Mutso et al. 2018).

Despite the continuing progress in elucidating the mechanism of alphavirus RNA replication, the critical steps of spherule formation and function remain poorly understood. As described above, these fascinating membrane structures (Fig. 4) are initially formed at the plasma membrane and, at later times post-infection, are translocated into the cytoplasm as part of the CPV-I membranes. For SINV, VEEV, and CHIKV, this translocation is incomplete and large numbers of RCs remain at the plasma membrane of both vertebrate and mosquito cells. SFV exhibits some differences, because late in infection, almost the entire pool of spherules is transported into the cytoplasm (Spuul et al. 2010). This internalization is determined by the actin–myosin network and presence of active phosphatidylinositol 3-kinases (PI3Ks). Inhibition of microtubule polymerization by nocodazole has a negative effect on the rates of viral RNA synthesis and a stronger effect on the rate of infectious virus release (Frolova et al. 2010; Spuul et al. 2010). However, inhibition of PI3Ks stalls SINV- and SFV-specific spherules at the plasma membrane without a detectable effect on viral replication rates (Frolova et al. 2010; Spuul et al. 2010). PI3K is a key component of the PI3K–Akt–mammalian target of rapamycin (mTOR) pathway, and this pathway is strongly activated during SFV infection, but CHIKV replication, in contrast, activates it relatively inefficiently (Thaa et al. 2015). This provides a plausible explanation for the detected differences in internalization of SFV and CHIKV spherules from the plasma membrane to the cytoplasm.

5 Conclusions

Thus, the molecular mechanism of RC and spherule formation, and exact functions of viral and host factors in these processes remain insufficiently understood. To date, the enzymatic activities of alphavirus nsPs have been reasonably well established. However, the overall structure of RC, its organization, and the exact place of RNA synthesis in the spherules remain obscure. The most poorly understood processes include (i) functions of numerous host factors in RC formation and RNA replication, (ii) the recruitment of viral G RNA into RC, (iii) coordination of different nsPs and promoters of G RNA and SG RNA in RC function, (iv) movement of dsRNA, which is packed inside the spherule cavity, at least relative to nsP4-associated RdRp, (v) the mechanism of newly synthesized ssRNA release from spherule-associated RC. This list can be continued. Taken together, the accumulated data suggest that we are only beginning to understand the mechanism of replication for these small, relatively simple, enveloped viruses and functions of the encoded handful of nonstructural and structural proteins.

References

Abraham R, Hauer D, McPherson RL, Utt A, Kirby IT, Cohen MS, Merits A, Leung AKL, Griffin DE (2018) ADP-ribosyl-binding and hydrolase activities of the alphavirus nsP3 macrodomain are critical for initiation of virus replication. Proc Natl Acad Sci U S A 115 (44): E10457–E10466. https://doi.org/10.1073/pnas.1812130115

Ahola T, Kaariainen L (1995) Reaction in alphavirus mRNA capping: formation of a covalent complex of nonstructural protein nsP1 with 7-methyl-GMP. Proc Natl Acad Sci U S A 92 (2):507–511

Ahola T, Laakkonen P, Vihinen H, Kaariainen L (1997) Critical residues of Semliki Forest virus RNA capping enzyme involved in methyltransferase and guanylyltransferase-like activities. J Virol 71(1):392–397

Akhrymuk I, Frolov I, Frolova EI (2016) Both RIG-I and MDA5 detect alphavirus replication in concentration-dependent mode. Virology 487:230–241. https://doi.org/10.1016/j.virol.2015.09.023

Akhrymuk I, Frolov I, Frolova EI (2018a) Sindbis Virus Infection Causes Cell Death by nsP2-Induced Transcriptional Shutoff or by nsP3-Dependent Translational Shutoff. J Virol 92 (23):e01388. https://doi.org/10.1128/JVI.01388-18

Akhrymuk I, Kulemzin SV, Frolova EI (2012) Evasion of the innate immune response: the Old World alphavirus nsP2 protein induces rapid degradation of Rpb1, a catalytic subunit of RNA polymerase II. J Virol 86(13):7180–7191. https://doi.org/10.1128/JVI.00541-12

Akhrymuk I, Lukash T, Frolov I, Frolova EI (2018b) Novel mutations in nsP2 abolish chikungunya virus-induced transcriptional shutoff and make virus less cytopathic without affecting its replication rates. J Virol. https://doi.org/10.1128/jvi.02062-18

Atasheva S, Gorchakov R, English R, Frolov I, Frolova E (2007) Development of Sindbis viruses encoding nsP2/GFP chimeric proteins and their application for studying nsP2 functioning. J Virol 81(10):5046–5057. https://doi.org/10.1128/JVI.02746-06 (JVI.02746-06 [pii])

Blair CD (2011) Mosquito RNAi is the major innate immune pathway controlling arbovirus infection and transmission. Future Microbiol 6(3):265–277. https://doi.org/10.2217/fmb.11.11

Chen R, Wang E, Tsetsarkin KA, Weaver SC (2013) Chikungunya virus 3′ untranslated region: adaptation to mosquitoes and a population bottleneck as major evolutionary forces. PLoS Pathog 9(8):e1003591. https://doi.org/10.1371/journal.ppat.1003591 (PPATHOGENS-D-13-01023 [pii])

Cristea IM, Carroll JW, Rout MP, Rice CM, Chait BT, MacDonald MR (2006) Tracking and elucidating alphavirus-host protein interactions. J Biol Chem 281(40):30269–30278. https://doi.org/10.1074/jbc.M603980200

Cristea IM, Rozjabek H, Molloy KR, Karki S, White LL, Rice CM, Rout MP, Chait BT, MacDonald MR (2010) Host factors associated with the Sindbis virus RNA-dependent RNA polymerase: role for G3BP1 and G3BP2 in virus replication. J Virol 84(13):6720–6732. https://doi.org/10.1128/JVI.01983-09

Das PK, Merits A, Lulla A (2014) Functional cross-talk between distant domains of chikungunya virus non-structural protein 2 is decisive for its RNA-modulating activity. J Biol Chem 289 (9):5635–5653. https://doi.org/10.1074/jbc.M113.503433

de Groot RJ, Rümenapf T, Kuhn RJ, Strauss EG, Strauss JH (1991) Sindbis virus RNA polymerase is degraded by the N-end rule pathway. Proc Natl Acad Sci USA 88:8967–8971

Dryga SA, Dryga OA, Schlesinger S (1997) Identification of mutations in a Sindbis virus variant able to establish persistent infection in BHK cells: The importance of a mutation in the nsP2 gene. Virology 228:72–83

Eckei L, Krieg S, Butepage M, Lehmann A, Gross A, Lippok B, Grimm AR, Kummerer BM, Rossetti G, Luscher B, Verheugd P (2017) The conserved macrodomains of the non-structural proteins of Chikungunya virus and other pathogenic positive strand RNA viruses function as mono-ADP-ribosylhydrolases. Sci Rep 7:41746. https://doi.org/10.1038/srep41746

Egloff MP, Malet H, Putics A, Heinonen M, Dutartre H, Frangeul A, Gruez A, Campanacci V, Cambillau C, Ziebuhr J, Ahola T, Canard B (2006) Structural and functional basis for ADP-ribose and poly(ADP-ribose) binding by viral macro domains. J Virol 80(17):8493–8502. https://doi.org/10.1128/JVI.00713-06 (80/17/8493 [pii])

Fayzulin R, Frolov I (2004) Changes of the secondary structure of the 5′ end of the Sindbis virus genome inhibit virus growth in mosquito cells and lead to accumulation of adaptive mutations. J Virol 78(10):4953–4964

Foy NJ, Akhrymuk M, Akhrymuk I, Atasheva S, Bopda-Waffo A, Frolov I, Frolova EI (2013a) Hypervariable domains of nsP3 proteins of New World and Old World alphaviruses mediate formation of distinct, virus-specific protein complexes. J Virol 87(4):1997–2010. https://doi.org/10.1128/JVI.02853-12

Foy NJ, Akhrymuk M, Shustov AV, Frolova EI, Frolov I (2013b) Hypervariable domain of nonstructural protein nsP3 of Venezuelan equine encephalitis virus determines cell-specific mode of virus replication. J Virol 87(13):7569–7584. https://doi.org/10.1128/JVI.00720-13

Friedman RM, Levin JG, Grimley PM, Berezesky IK (1972) Membrane-associated replication complex in arbovirus infection. J Virol 10(3):504–515

Frolov I, Agapov E, Hoffman TA Jr, Prágai BM, Lippa M, Schlesinger S, Rice CM (1999) Selection of RNA replicons capable of persistent noncytopathic replication in mammalian cells. J Virol 73:3854–3865

Frolov I, Akhrymuk M, Akhrymuk I, Atasheva S, Frolova EI (2012) Early events in alphavirus replication determine the outcome of infection. J Virol 86(9):5055–5066. https://doi.org/10.1128/JVI.07223-11

Frolov I, Garmashova N, Atasheva S, Frolova EI (2009) Random insertion mutagenesis of sindbis virus nonstructural protein 2 and selection of variants incapable of downregulating cellular transcription. J Virol 83(18):9031–9044. https://doi.org/10.1128/JVI.00850-09

Frolov I, Hardy R, Rice CM (2001) Cis-acting RNA elements at the 5′ end of Sindbis virus genome RNA regulate minus- and plus-strand RNA synthesis. RNA 7(11):1638–1651

Frolov I, Hoffman TA, Prágai BM, Dryga SA, Huang HV, Schlesinger S, Rice CM (1996) Alphavirus-based expression systems: strategies and applications. Proc Natl Acad Sci U S A 93:11371–11377

Frolov I, Kim DY, Akhrymuk M, Mobley JA, Frolova EI (2017) Hypervariable Domain of Eastern Equine Encephalitis Virus nsP3 Redundantly Utilizes Multiple Cellular Proteins for Replication Complex Assembly. J Virol 91(14):e00371–00317. https://doi.org/10.1128/jvi.00371-17

Frolov I, Schlesinger S (1994) Comparison of the effects of Sindbis virus and Sindbis virus replicons on host cell protein synthesis and cytopathogenicity in BHK cells. J Virol 68:1721–1727

Frolova E, Frolov I, Schlesinger S (1997) Packaging signals in alphaviruses. J Virol 71(1):248–258

Frolova E, Gorchakov R, Garmashova N, Atasheva S, Vergara LA, Frolov I (2006) Formation of nsP3-specific protein complexes during Sindbis virus replication. J Virol 80(8):4122–4134. https://doi.org/10.1128/JVI.80.8.4122-4134.2006

Frolova EI, Fayzulin RZ, Cook SH, Griffin DE, Rice CM, Frolov I (2002) Roles of nonstructural protein nsP2 and Alpha/Beta interferons in determining the outcome of Sindbis virus infection. J Virol 76(22):11254–11264

Frolova EI, Gorchakov R, Pereboeva L, Atasheva S, Frolov I (2010) Functional Sindbis virus replicative complexes are formed at the plasma membrane. J Virol 84(22):11679–11695. https://doi.org/10.1128/JVI.01441-10

Fros JJ, Domeradzka NE, Baggen J, Geertsema C, Flipse J, Vlak JM, Pijlman GP (2012) Chikungunya virus nsP3 blocks stress granule assembly by recruitment of G3BP into cytoplasmic foci. J Virol 86(19):10873–10879. https://doi.org/10.1128/JVI.01506-12

Fros JJ, Geertsema C, Zouache K, Baggen J, Domeradzka N, van Leeuwen DM, Flipse J, Vlak JM, Failloux AB, Pijlman GP (2015) Mosquito Rasputin interacts with chikungunya virus nsP3 and determines the infection rate in Aedes albopictus. Parasit Vectors 8:464. https://doi.org/10.1186/s13071-015-1070-4 (1186/s13071-015-1070-4 [pii])

Fros JJ, Liu WJ, Prow NA, Geertsema C, Ligtenberg M, Vanlandingham DL, Schnettler E, Vlak JM, Suhrbier A, Khromykh AA, Pijlman GP (2010) Chikungunya virus nonstructural protein 2 inhibits type I/II interferon-stimulated JAK-STAT signaling. J Virol 84(20):10877–10887. https://doi.org/10.1128/JVI.00949-10

Fros JJ, van der Maten E, Vlak JM, Pijlman GP (2013) The C-terminal domain of chikungunya virus nsP2 independently governs viral RNA replication, cytopathicity, and inhibition of interferon signaling. J Virol 87(18):10394–10400. https://doi.org/10.1128/JVI.00884-13

Froshauer S, Kartenbeck J, Helenius A (1988) Alphavirus RNA replicase is located on the cytoplasmic surface of endosomes and lysosomes. J Cell Biol 107(6 Pt 1):2075–2086

Garmashova N, Gorchakov R, Frolova E, Frolov I (2006) Sindbis virus nonstructural protein nsP2 is cytotoxic and inhibits cellular transcription. J Virol 80(12):5686–5696. https://doi.org/10.1128/JVI.02739-05

Garmashova N, Gorchakov R, Volkova E, Paessler S, Frolova E, Frolov I (2007) The Old World and New World alphaviruses use different virus-specific proteins for induction of transcriptional shutoff. J Virol 81(5):2472–2484. https://doi.org/10.1128/JVI.02073-06

Gomez de Cedron M, Ehsani N, Mikkola ML, Garcia JA, Kaariainen L (1999) RNA helicase activity of Semliki Forest virus replicase protein NSP2. FEBS Lett 448(1):19–22

Gorbalenya AE, Koonin EV, Donchenko AP, Blinov VM (1989) Two related superfamilies of putative helicases involved in replication, recombination, repair and expression of DNA and RNA genomes. Nucleic Acids Res 17(12):4713–4729

Gorchakov R, Frolova E, Sawicki S, Atasheva S, Sawicki D, Frolov I (2008a) A new role for ns polyprotein cleavage in Sindbis virus replication. J Virol 82(13):6218–6231. https://doi.org/10.1128/JVI.02624-07

Gorchakov R, Garmashova N, Frolova E, Frolov I (2008b) Different types of nsP3-containing protein complexes in Sindbis virus-infected cells. J Virol 82(20):10088–10101. https://doi.org/10.1128/JVI.01011-08

Gorchakov R, Hardy R, Rice CM, Frolov I (2004) Selection of functional 5′ cis-acting elements promoting efficient sindbis virus genome replication. J Virol 78(1):61–75

Grakoui A, Levis R, Raju R, Huang HV, Rice CM (1989) A *cis*-acting mutation in the Sindbis virus junction region which affects subgenomic RNA synthesis. J Virol 63:5216–5227

Grimley PM, Berezesky IK, Friedman RM (1968) Cytoplasmic structures associated with an arbovirus infection: loci of viral ribonucleic acid synthesis. J Virol 2(11):1326–1338

Grimley PM, Levin JG, Berezesky IK, Friedman RM (1972) Specific membranous structures associated with the replication of Group A arboviruses. J Virol 10:492–503

Hallengard D, Kakoulidou M, Lulla A, Kummerer BM, Johansson DX, Mutso M, Lulla V, Fazakerley JK, Roques P, Le Grand R, Merits A, Liljestrom P (2014) Novel attenuated Chikungunya vaccine candidates elicit protective immunity in C57BL/6 mice. J Virol 88(5):2858–2866. https://doi.org/10.1128/JVI.03453-13 (JVI.03453-13 [pii])

Hardy WR, Strauss JH (1989) Processing the nonstructural proteins of Sindbis virus: nonstructural proteinase is in the C-terminal half of nsP2 and functions both in *cis* and *trans*. J Virol 63:4653–4664

Hyde JL, Gardner CL, Kimura T, White JP, Liu G, Trobaugh DW, Huang C, Tonelli M, Paessler S, Takeda K, Klimstra WB, Amarasinghe GK, Diamond MS (2014) A viral RNA structural element alters host recognition of nonself RNA. Science 343(6172):783–787. https://doi.org/10.1126/science.1248465

Jones PH, Maric M, Madison MN, Maury W, Roller RJ, Okeoma CM (2013) BST-2/tetherin-mediated restriction of chikungunya (CHIKV) VLP budding is counteracted by CHIKV non-structural protein 1 (nsP1). Virology 438(1):37–49. https://doi.org/10.1016/j.virol.2013.01.010

Kallio K, Hellstrom K, Balistreri G, Spuul P, Jokitalo E, Ahola T (2013) Template RNA length determines the size of replication complex spherules for Semliki Forest virus. J Virol 87(16):9125–9134. https://doi.org/10.1128/JVI.00660-13

Karpe YA, Aher PP, Lole KS (2011) NTPase and 5′-RNA triphosphatase activities of Chikungunya virus nsP2 protein. PLoS ONE 6(7):e22336. https://doi.org/10.1371/journal.pone.0022336 (PONE-D-11-04866 [pii])

Khan AH, Morita K, Parquet Md Mdel C, Hasebe F, Mathenge EG, Igarashi A (2002) Complete nucleotide sequence of chikungunya virus and evidence for an internal polyadenylation site. J Gen Virol 83(Pt 12):3075–3084. https://doi.org/10.1099/0022-1317-83-12-3075

Kim DY, Atasheva S, Foy NJ, Wang E, Frolova EI, Weaver S, Frolov I (2011a) Design of chimeric alphaviruses with a programmed, attenuated, cell type-restricted phenotype. J Virol 85(9):4363–4376. https://doi.org/10.1128/JVI.00065-11

Kim DY, Firth AE, Atasheva S, Frolova EI, Frolov I (2011b) Conservation of a packaging signal and the viral genome RNA packaging mechanism in alphavirus evolution. J Virol 85 (16):8022–8036. https://doi.org/10.1128/JVI.00644-11

Kim DY, Reynaud JM, Rasalouskaya A, Akhrymuk I, Mobley JA, Frolov I, Frolova EI (2016) New World and old world alphaviruses have evolved to exploit different components of stress granules, FXR and G3BP proteins, for assembly of viral replication complexes. PLoS Pathog 12(8):e1005810. https://doi.org/10.1371/journal.ppat.1005810

Kim KH, Rumenapf T, Strauss EG, Strauss JH (2004) Regulation of Semliki Forest virus RNA replication: a model for the control of alphavirus pathogenesis in invertebrate hosts. Virology 323(1):153–163

Kristensen O (2015) Crystal structure of the G3BP2 NTF2-like domain in complex with a canonical FGDF motif peptide. Biochem Biophys Res Commun 467(1):53–57. https://doi.org/10.1016/j.bbrc.2015.09.123 (S0006-291X(15)30640-9 [pii])

Kuhn RJ, Hong Z, Strauss JH (1990) Mutagenesis of 3′ nontranslated region of Sindbis virus RNA. J Virol 64:1465–1476

Kuhn RJ, Niesters HGM, Hong Z, Strauss JH (1991) Infectious RNA transcripts from Ross River virus cDNA clones and the construction and characterization of defined chimeras with Sindbis virus. Virology 182:430–441

Kujala P, Ikaheimonen A, Ehsani N, Vihinen H, Auvinen P, Kaariainen L (2001) Biogenesis of the Semliki Forest virus RNA replication complex. J Virol 75(8):3873–3884

Kujala P, Rikkonen M, Ahola T, Kelve M, Saarma M, Kaariainen L (1997) Monoclonal antibodies specific for Semliki Forest virus replicase protein nsP2. J Gen Virol 78(Pt 2):343–351

Kulasegaran-Shylini R, Atasheva S, Gorenstein DG, Frolov I (2009a) Structural and functional elements of the promoter encoded by the 5′ untranslated region of the Venezuelan equine encephalitis virus genome. J Virol 83(17):8327–8339

Kulasegaran-Shylini R, Thiviyanathan V, Gorenstein DG, Frolov I (2009b) The 5′UTR-specific mutation in VEEV TC-83 genome has a strong effect on RNA replication and subgenomic RNA synthesis, but not on translation of the encoded proteins. Virology 387(1):211–221

Kumar P, Sweeney TR, Skabkin MA, Skabkina OV, Hellen CU, Pestova TV (2014) Inhibition of translation by IFIT family members is determined by their ability to interact selectively with the 5′-terminal regions of cap0-, cap1- and 5′ppp- mRNAs. Nucleic Acids Res 42(5):3228–3245. https://doi.org/10.1093/nar/gkt1321 (gkt1321 [pii])

Laakkonen P, Auvinen P, Kujala P, Kaariainen L (1998) Alphavirus replicase protein NSP1 induces filopodia and rearrangement of actin filaments. J Virol 72(12):10265–10269

Langsjoen RM, Rubinstein RJ, Kautz TF, Auguste AJ, Erasmus JH, Kiaty-Figueroa L, Gerhardt R, Lin D, Hari KL, Jain R, Ruiz N, Muruato AE, Silfa J, Bido F, Dacso M, Weaver SC (2016) Molecular virologic and clinical characteristics of a Chikungunya Fever Outbreak in La Romana, Dominican Republic, 2014. PLoS Negl Trop Dis 10(12):e0005189. https://doi.org/10.1371/journal.pntd.0005189 (PNTD-D-16-00891 [pii])

LaStarza MW, Grakoui A, Rice CM (1994) Deletion and duplication mutations in the C-terminal nonconserved region of Sindbis virus: effects on phosphorylation and on virus replication in vertebrate and invertebrate cells. Virology 202:224–232

Lehtovaara P, Söderlund H, Keränen S, Pettersson RF, Kääriäinen L (1981) 18S defective interfering RNA of Semliki Forest virus contains a triplicated linear repeat. Proc Natl Acad Sci USA 78:5353-5357

Lemm JA, Rice CM (1993a) Assembly of functional Sindbis virus RNA replication complexes: requirement for coexpression of P123 and P34. J Virol 67:1905–1915

Lemm JA, Rice CM (1993b) Roles of nonstructural polyproteins and cleavage products in regulating Sindbis virus RNA replication and transcription. J Virol 67:1916–1926

Lemm JA, Rümenapf T, Strauss EG, Strauss JH, Rice CM (1994) Polypeptide requirements for assembly of functional Sindbis virus replication complexes: a model for the temporal regulation of minus and plus-strand RNA synthesis. EMBO J 13:2925–2934

Liljestrom P (1994) Alphavirus expression systems. Curr Opin Biotechnol 5(5):495–500

Liljeström P, Garoff H (1991) A new generation of animal cell expression vectors based on the Semliki Forest virus replicon. Biotechnology 9:1356–1361

Lulla A, Lulla V, Merits A (2012) Macromolecular assembly-driven processing of the 2/3 cleavage site in the alphavirus replicase polyprotein. J Virol 86(1):553–565. https://doi.org/10.1128/JVI.05195-11

Lulla V, Karo-Astover L, Rausalu K, Merits A, Lulla A (2013) Presentation overrides specificity: probing the plasticity of alphaviral proteolytic activity through mutational analysis. J Virol 87 (18):10207–10220. https://doi.org/10.1128/JVI.01485-13

Lulla V, Merits A, Sarin P, Kaariainen L, Keranen S, Ahola T (2006) Identification of mutations causing temperature-sensitive defects in Semliki Forest virus RNA synthesis. J Virol 80 (6):3108–3111. https://doi.org/10.1128/JVI.80.6.3108-3111.2006 (80/6/3108 [pii])

Mai J, Sawicki SG, Sawicki DL (2009) Fate of minus-strand templates and replication complexes produced by a p23-cleavage-defective mutant of Sindbis virus. J Virol 83(17):8553–8564. https://doi.org/10.1128/JVI.00056-09

Malet H, Coutard B, Jamal S, Dutartre H, Papageorgiou N, Neuvonen M, Ahola T, Forrester N, Gould EA, Lafitte D, Ferron F, Lescar J, Gorbalenya AE, de Lamballerie X, Canard B (2009) The crystal structures of Chikungunya and Venezuelan equine encephalitis virus nsP3 macro domains define a conserved adenosine binding pocket. J Virol 83(13):6534–6545. https://doi.org/10.1128/JVI.00189-09 (JVI.00189-09 [pii])

Mayuri Geders TW, Smith JL, Kuhn RJ (2008) Role for conserved residues of sindbis virus nonstructural protein 2 methyltransferase-like domain in regulation of minus-strand synthesis and development of cytopathic infection. J Virol 82(15):7284–7297

McInerney GM, Kedersha NL, Kaufman RJ, Anderson P, Liljestrom P (2005) Importance of eIF2alpha phosphorylation and stress granule assembly in alphavirus translation regulation. Mol Biol Cell 16(8):3753–3763. https://doi.org/10.1091/mbc.E05-02-0124

McPherson RL, Abraham R, Sreekumar E, Ong SE, Cheng SJ, Baxter VK, Kistemaker HA, Filippov DV, Griffin DE, Leung AK (2017) ADP-ribosylhydrolase activity of Chikungunya virus macrodomain is critical for virus replication and virulence. Proc Natl Acad Sci U S A 114 (7):1666–1671. https://doi.org/10.1073/pnas.1621485114

McSweegan E, Weaver SC, Lecuit M, Frieman M, Morrison TE, Hrynkow S (2015) The global virus network: challenging chikungunya. Antiviral Res 120:147–152. https://doi.org/10.1016/j.antiviral.2015.06.003 (S0166-3542(15)00134-5 [pii])

Meshram CD, Agback P, Shiliaev N, Urakova N, Mobley JA, Agback T, Frolova EI, Frolov I (2018) Multiple host factors interact with hypervariable domain of Chikungunya Virus nsP3 and Determine Viral Replication in Cell-Specific Mode. J Virol. https://doi.org/10.1128/jvi.00838-18

Michel G, Petrakova O, Atasheva S, Frolov I (2007) Adaptation of Venezuelan equine encephalitis virus lacking 51-nt conserved sequence element to replication in mammalian and mosquito cells. Virology 362(2):475–487. https://doi.org/10.1016/j.virol.2007.01.009

Monroe SS, Schlesinger S (1983) RNAs from two independently isolated defective interfering particles of Sindbis virus contain a cellular tRNA sequence at their 5′ ends. Proc Natl Acad Sci USA 80(11):3279-3283

Mounce BC, Cesaro T, Vlajnic L, Vidina A, Vallet T, Weger-Lucarelli J, Passoni G, Stapleford KA, Levraud JP, Vignuzzi M (2017) Chikungunya Virus Overcomes Polyamine Depletion by Mutation of nsP1 and the Opal Stop Codon To Confer Enhanced Replication and Fitness. J Virol 91(15). https://doi.org/10.1128/jvi.00344-17

Mutso M, Morro AM, Smedberg C, Kasvandik S, Aquilimeba M, Teppor M, Tarve L, Lulla A, Lulla V, Saul S, Thaa B, McInerney GM, Merits A, Varjak M (2018) Mutation of CD2AP and SH3KBP1 Binding Motif in Alphavirus nsP3 Hypervariable Domain Results in Attenuated Virus. Viruses 10(5). https://doi.org/10.3390/v10050226

Myles KM, Kelly CL, Ledermann JP, Powers AM (2006) Effects of an opal termination codon preceding the nsP4 gene sequence in the O'Nyong-Nyong virus genome on Anopheles gambiae infectivity. J Virol 80(10):4992–4997. https://doi.org/10.1128/JVI.80.10.4992-4997.2006

Nappe TM, Chuhran CM, Johnson SA (2016) The Chikungunya virus: an emerging US pathogen. World J Emerg Med 7(1):65–67. https://doi.org/10.5847/wjem.j.1920-8642.2016.01.012 (WJEM-7-65 [pii])

Nasar F, Palacios G, Gorchakov RV, Guzman H, Da Rosa AP, Savji N, Popov VL, Sherman MB, Lipkin WI, Tesh RB, Weaver SC (2012) Eilat virus, a unique alphavirus with host range restricted to insects by RNA replication. Proc Natl Acad Sci U S A 109(36):14622–14627. https://doi.org/10.1073/pnas.1204787109

Neuvonen M, Ahola T (2009) Differential activities of cellular and viral macro domain proteins in binding of ADP-ribose metabolites. J Mol Biol 385(1):212–225. https://doi.org/10.1016/j.jmb.2008.10.045 (S0022-2836(08)01334-X [pii])

Neuvonen M, Kazlauskas A, Martikainen M, Hinkkanen A, Ahola T, Saksela K (2011) SH3 domain-mediated recruitment of host cell amphiphysins by alphavirus nsP3 promotes viral RNA replication. PLoS Pathog 7(11):e1002383. https://doi.org/10.1371/journal.ppat.1002383 (PPATHOGENS-D-11-01401 [pii])

Niesters HGM, Strauss JH (1990) Defined mutations in the 5′ nontranslated sequence of Sindbis virus RNA. J Virol 64:4162–4168

Nikonov A, Molder T, Sikut R, Kiiver K, Mannik A, Toots U, Lulla A, Lulla V, Utt A, Merits A, Ustav M (2013) RIG-I and MDA-5 detection of viral RNA-dependent RNA polymerase activity restricts positive-strand RNA virus replication. PLoS Pathog 9(9):e1003610. https://doi.org/10.1371/journal.ppat.1003610

Ou JH, Trent DW, Strauss JH (1982) The 3′-non-coding regions of alphavirus RNAs contain repeating sequences. J Mol Biol 156(4):719–730

Panas MD, Ahola T, McInerney GM (2014) The C-terminal repeat domains of nsP3 from the Old World alphaviruses bind directly to G3BP. J Virol 88(10):5888–5893. https://doi.org/10.1128/JVI.00439-14

Panas MD, Varjak M, Lulla A, Eng KE, Merits A, Karlsson Hedestam GB, McInerney GM (2012) Sequestration of G3BP coupled with efficient translation inhibits stress granules in Semliki Forest virus infection. Mol Biol Cell 23(24):4701–4712. https://doi.org/10.1091/mbc.E12-08-0619

Park E, Griffin DE (2009) The nsP3 macro domain is important for Sindbis virus replication in neurons and neurovirulence in mice. Virology 388(2):305–314. https://doi.org/10.1016/j.virol.2009.03.031 (S0042-6822(09)00209-8 [pii])

Peranen J, Rikkonen M, Liljestrom P, Kaariainen L (1990) Nuclear localization of Semliki Forest virus-specific nonstructural protein nsP2. J Virol 64(5):1888–1896

Perri S, Driver DA, Gardner JP, Sherrill S, Belli BA, Dubensky TW Jr, Polo JM (2000) Replicon vectors derived from Sindbis virus and Semliki forest virus that establish persistent replication in host cells. J Virol 74(20):9802–9807

Plante K, Wang E, Partidos CD, Weger J, Gorchakov R, Tsetsarkin K, Borland EM, Powers AM, Seymour R, Stinchcomb DT, Osorio JE, Frolov I, Weaver SC (2011) Novel chikungunya vaccine candidate with an IRES-based attenuation and host range alteration mechanism. PLoS Pathog 7(7):e1002142. https://doi.org/10.1371/journal.ppat.1002142

Rack JG, Perina D, Ahel I (2016) Macrodomains: structure, function, evolution, and catalytic activities. Annu Rev Biochem 85:431–454. https://doi.org/10.1146/annurev-biochem-060815-014935

Raju R, Hajjou M, Hill KR, Botta V, Botta S (1999) In vivo addition of poly(A) tail and AU-rich sequences to the 3′ terminus of the Sindbis virus RNA genome: a novel 3′-end repair pathway. J Virol 73(3):2410–2419

Rausalu K, Utt A, Quirin T, Varghese FS, Zusinaite E, Das PK, Ahola T, Merits A (2016) Chikungunya virus infectivity, RNA replication and non-structural polyprotein processing depend on the nsP2 protease's active site cysteine residue. Sci Rep 6:37124. https://doi.org/10.1038/srep37124 (srep37124 [pii])

Reynaud JM, Kim DY, Atasheva S, Rasalouskaya A, White JP, Diamond MS, Weaver SC, Frolova EI, Frolov I (2015) IFIT1 differentially interferes with translation and replication of alphavirus genomes and promotes induction of type I interferon. PLoS Pathog 11(4):e1004863. https://doi.org/10.1371/journal.ppat.1004863

Rikkonen M, Peranen J, Kaariainen L (1992) Nuclear and nucleolar targeting signals of Semliki Forest virus nonstructural protein nsP2. Virology 189(2):462–473

Rikkonen M, Peranen J, Kaariainen L (1994) ATPase and GTPase activities associated with Semliki Forest virus nonstructural protein nsP2. J Virol 68(9):5804–5810

Robinson MC (1955) An epidemic of virus disease in Southern Province, Tanganyika Territory, in 1952-53. Trans Roy S Trop Med Hyg 49:28

Rubach JK, Wasik BR, Rupp JC, Kuhn RJ, Hardy RW, Smith JL (2009) Characterization of purified Sindbis virus nsP4 RNA-dependent RNA polymerase activity in vitro. Virology 384 (1):201–208

Rupp JC, Jundt N, Hardy RW (2011) Requirement for the amino-terminal domain of sindbis virus nsP4 during virus infection. J Virol 85(7):3449–3460. https://doi.org/10.1128/JVI.02058-10

Rupp JC, Sokoloski KJ, Gebhart NN, Hardy RW (2015) Alphavirus RNA synthesis and non-structural protein functions. J Gen Virol 96(9):2483–2500. https://doi.org/10.1099/jgv.0.000249

Russo AT, White MA, Watowich SJ (2006) The crystal structure of the Venezuelan equine encephalitis alphavirus nsP2 protease. Structure 14(9):1449–1458. https://doi.org/10.1016/j.str.2006.07.010 (S0969-2126(06)00333-9 [pii])

Sawicki DL, Gomatos PJ (1976) Replication of semliki forest virus: polyadenylate in plus-strand RNA and polyuridylate in minus-strand RNA. J Virol 20(2):446–464

Scholte FE, Tas A, Albulescu IC, Zusinaite E, Merits A, Snijder EJ, van Hemert MJ (2015) Stress granule components G3BP1 and G3BP2 play a proviral role early in Chikungunya virus replication. J Virol 89(8):4457–4469. https://doi.org/10.1128/JVI.03612-14

Scholte FE, Tas A, Martina BE, Cordioli P, Narayanan K, Makino S, Snijder EJ, van Hemert MJ (2013) Characterization of synthetic Chikungunya viruses based on the consensus sequence of recent E1-226 V isolates. PLoS ONE 8(8):e71047. https://doi.org/10.1371/journal.pone.0071047

Shin G, Yost SA, Miller MT, Elrod EJ, Grakoui A, Marcotrigiano J (2012) Structural and functional insights into alphavirus polyprotein processing and pathogenesis. Proc Natl Acad Sci U S A 109(41):16534–16539. https://doi.org/10.1073/pnas.1210418109

Shirako Y, Strauss JH (1990) Cleavage between nsP1 and nsP2 initiates the processing pathway of Sindbis virus nonstructural polyprotein P123. Virology 177:54–64

Shirako Y, Strauss JH (1994) Regulation of Sindbis virus RNA replication: uncleaved P123 and nsP4 function in minus strand RNA synthesis whereas cleaved products from P123 are required for efficient plus strand RNA synthesis. J Virol 185:1874–1885

Siomi MC, Eder PS, Kataoka N, Wan L, Liu Q, Dreyfuss G (1997) Transportin-mediated nuclear import of heterogeneous nuclear RNP proteins. J Cell Biol 138(6):1181–1192

Sokoloski KJ, Dickson AM, Chaskey EL, Garneau NL, Wilusz CJ, Wilusz J (2010) Sindbis virus usurps the cellular HuR protein to stabilize its transcripts and promote productive infections in mammalian and mosquito cells. Cell Host Microbe 8(2):196–207. https://doi.org/10.1016/j.chom.2010.07.003 (S1931-3128(10)00242-8 [pii])

Sokoloski KJ, Haist KC, Morrison TE, Mukhopadhyay S, Hardy RW (2015) Noncapped alphavirus genomic RNAs and their role during infection. J Virol 89(11):6080–6092. https://doi.org/10.1128/JVI.00553-15

Sokoloski KJ, Nease LM, May NA, Gebhart NN, Jones CE, Morrison TE, Hardy RW (2017) Identification of interactions between Sindbis virus capsid protein and cytoplasmic vRNA as novel virulence determinants. PLoS Pathog 13(6):e1006473. https://doi.org/10.1371/journal.ppat.1006473 (PPATHOGENS-D-16-02887 [pii])

Spuul P, Balistreri G, Kaariainen L, Ahola T (2010) Phosphatidylinositol 3-kinase-, actin-, and microtubule-dependent transport of Semliki Forest Virus replication complexes from the plasma membrane to modified lysosomes. J Virol 84(15):7543–7557. https://doi.org/10.1128/JVI.00477-10 (JVI.00477-10 [pii])

Strauss EG, deGroot RJ, Levinson R, Strauss JH (1992) Identification of the active site residues in the nsP2 proteinase of Sindbis virus. Virology 191:932–940

Strauss EG, Rice CM, Strauss JH (1983) Sequence coding for the alphavirus nonstructural proteins is interrupted by an opal termination codon. Proc Natl Acad Sci U S A 80(17):5271–5275

Strauss JH, Strauss EG (1994) The alphaviruses: gene expression, replication, evolution. Microbiol Rev 58:491–562

Sun C, Gardner CL, Watson AM, Ryman KD, Klimstra WB (2014) Stable, high-level expression of reporter proteins from improved alphavirus expression vectors to track replication and dissemination during encephalitic and arthritogenic disease. J Virol 88(4):2035–2046. https://doi.org/10.1128/JVI.02990-13

Svejstrup JQ (2003) Rescue of arrested RNA polymerase II complexes. J Cell Sci 116(Pt 3):447–451

Takkinen K (1986) Complete nucleotide sequence of the nonstructural protein genes of Semliki Forest virus. Nucleic Acids Res 14(14):5667–5682

Tamberg N, Lulla V, Fragkoudis R, Lulla A, Fazakerley JK, Merits A (2007) Insertion of EGFP into the replicase gene of Semliki Forest virus results in a novel, genetically stable marker virus. J Gen Virol 88(Pt 4):1225–1230

Thaa B, Biasiotto R, Eng K, Neuvonen M, Gotte B, Rheinemann L, Mutso M, Utt A, Varghese F, Balistreri G, Merits A, Ahola T, McInerney GM (2015) Differential Phosphatidylinositol-3-Kinase-Akt-mTOR Activation by Semliki Forest and chikungunya viruses is dependent on nsP3 and connected to replication complex internalization. J Virol 89(22):11420–11437. https://doi.org/10.1128/JVI.01579-15

Tomar S, Hardy RW, Smith JL, Kuhn RJ (2006) Catalytic core of alphavirus nonstructural protein nsP4 possesses terminal adenylyltransferase activity. J Virol 80(20):9962–9969

Tossavainen H, Aitio O, Hellman M, Saksela K, Permi P (2016) Structural basis of the high affinity interaction between the Alphavirus Nonstructural Protein-3 (nsP3) and the SH3 domain of Amphiphysin-2. J Biol Chem 291(31):16307–16317. https://doi.org/10.1074/jbc.M116.732412

Trobaugh DW, Gardner CL, Sun C, Haddow AD, Wang E, Chapnik E, Mildner A, Weaver SC, Ryman KD, Klimstra WB (2014) RNA viruses can hijack vertebrate microRNAs to suppress innate immunity. Nature 506(7487):245–248. https://doi.org/10.1038/nature12869

Tsetsarkin KA, Vanlandingham DL, McGee CE, Higgs S (2007) A single mutation in chikungunya virus affects vector specificity and epidemic potential. PLoS Pathog 3(12):e201. https://doi.org/10.1371/journal.ppat.0030201

Utt A, Das PK, Varjak M, Lulla V, Lulla A, Merits A (2015) Mutations conferring a noncytotoxic phenotype on chikungunya virus replicons compromise enzymatic properties of nonstructural protein 2. J Virol 89(6):3145–3162. https://doi.org/10.1128/JVI.03213-14

Vasiljeva L, Merits A, Auvinen P, Kaariainen L (2000) Identification of a novel function of the alphavirus capping apparatus. RNA 5′-triphosphatase activity of Nsp2. J Biol Chem 275(23):17281–17287. https://doi.org/10.1074/jbc.M910340199 (M910340199 [pii])

Vasiljeva L, Merits A, Golubtsov A, Sizemskaja V, Kaariainen L, Ahola T (2003) Regulation of the sequential processing of Semliki Forest virus replicase polyprotein. J Biol Chem 278(43):41636–41645

Vihinen H, Ahola T, Tuittila M, Merits A, Kaariainen L (2001) Elimination of phosphorylation sites of Semliki Forest virus replicase protein nsP3. J Biol Chem 276(8):5745–5752

Volkova E, Frolova E, Darwin JR, Forrester NL, Weaver SC, Frolov I (2008) IRES-dependent replication of Venezuelan equine encephalitis virus makes it highly attenuated and incapable of replicating in mosquito cells. Virology 377(1):160–169. https://doi.org/10.1016/j.virol.2008.04.020

Wang E, Kim DY, Weaver SC, Frolov I (2011) Chimeric Chikungunya viruses are nonpathogenic in highly sensitive mouse models but efficiently induce a protective immune response. J Virol 85(17):9249–9252. https://doi.org/10.1128/JVI.00844-11

Wang E, Volkova E, Adams AP, Forrester N, Xiao SY, Frolov I, Weaver SC (2008) Chimeric alphavirus vaccine candidates for chikungunya. Vaccine 26(39):5030–5039

Weaver SC, Forrester NL (2015) Chikungunya: evolutionary history and recent epidemic spread. Antiviral Res 120:32–39. https://doi.org/10.1016/j.antiviral.2015.04.016

Weaver SC, Lecuit M (2015a) Chikungunya virus and the global spread of a mosquito-borne disease. N Engl J Med 372(13):1231–1239. https://doi.org/10.1056/NEJMra1406035

Weaver SC, Lecuit M (2015b) Chikungunya virus infections. N Engl J Med 373(1):94–95. https://doi.org/10.1056/NEJMc1505501

Weiss B, Geigenmuller-Gnirke U, Schlesinger S (1994) Interactions between Sindbis virus RNAs and a 68 amino acid derivative of the viral capsid protein further defines the capsid binding site. Nucleic Acids Res 22(5):780–786

Weiss B, Nitschko H, Ghattas I, Wright R, Schlesinger S (1989) Evidence for specificity in the encapsidation of Sindbis virus RNAs. J Virol 63:5310–5318

White CL, Thomson M, Dimmock NJ (1998) Deletion analysis of a defective interfering Semliki Forest virus RNA genome defines a region in the nsP2 sequence that is required for efficient packaging of the genome into virus particles. J Virol 72(5):4320–4326

White LK, Sali T, Alvarado D, Gatti E, Pierre P, Streblow D, Defilippis VR (2011) Chikungunya virus induces IPS-1-dependent innate immune activation and protein kinase R-independent translational shutoff. J Virol 85(1):606–620. https://doi.org/10.1128/JVI.00767-10 (JVI.00767-10 [pii])

Understanding Molecular Pathogenesis with Chikungunya Virus Research Tools

Guillaume Carissimo and Lisa F. P. Ng

Contents

Abstract Since its re-emergence in 2006, Chikungunya has been a major health concern in endemic areas. Transmitted by *Aedes* mosquitoes to mammalian hosts, Chikungunya leads to persistent debilitating symptoms in a high proportion of symptomatic human cases. In this review, we present several tools on the mosquito vector side as well as on the mammalian side that have been used to advance research on Chikungunya transmission and immunopathogenesis. These tools lead to key understandings of viral replication in both hosts, and innate and adaptive

G. Carissimo · L. F. P. Ng
Singapore Immunology Network (SIgN), Agency for Science, Technology and Research (A*STAR), Biopolis, Singapore, Singapore

L. F. P. Ng
Institute of Infection and Global Health, University of Liverpool, Liverpool, UK

L. F. P. Ng (✉)
Laboratory of Microbial Immunity, Singapore Immunology Network, 8A Biomedical Grove, #04-06, Immunos, Biopolis 138648, Singapore
e-mail: lisa_ng@immunol.a-star.edu.sg

Current Topics in Microbiology and Immunology (2022) 435: 33–53
https://doi.org/10.1007/82_2019_158

Published online: 20 March 2019

responses mediating virus clearance and pathology in mammals. This comprehension of viral mechanisms has allowed the development of promising treatment avenues in animal models that will need to be further explored. However, research efforts need to continue in order to develop better and unbiased tools to assess antiviral and treatment strategies as well as further understand immune mechanisms at play in human pathologies.

1 Chikungunya Virus and Disease

Chikungunya virus (CHIKV) is a virus of the genus *Alphaviridae* in the *Togaviridae* family. The first outbreak of the modern era was identified in Tanzania in 1952, where the indigenous Swahili population called the disease Chikungunya, meaning, "that which bends up" (Lumsden 1955; Robinson 1955). Retrospective studies of outbreaks previously described as dengue virus infections due to similar clinical symptoms now attributed to CHIKV, suggest that outbreaks were reported as early as 1779 (Carey 1971; Halstead 2015). Since 1952, CHIKV has circulated sporadically in Asia and Africa [reviewed in (Weaver and Forrester 2015)]. The worldwide epidemic originated from the local re-emergence in Kenya in 2004 before the outbreak in Indian Ocean islands including La Reunion in 2005–2006 (Cordel et al. 2006; Renault et al. 2007). Subsequently, CHIKV has caused massive outbreaks in several countries: Africa, the Indian Ocean, India, Asia, and Southeast Asia (Josseran et al. 2006; Kalantri et al. 2006; Hapuarachchi et al. 2010), Caribbean islands, North, Central, and South America (Leparc-Goffart et al. 2014; Van Bortel et al. 2014; Nunes et al. 2015), and even in temperate regions such as in Italy and Europe (Rezza et al. 2007).

Genetic analysis of isolated viruses has allowed classification of CHIKV into three distinct genotypes: West African, East-Central-South-African (ECSA), and the Asian genotypes. The strain responsible for the re-emergence of CHIKV originating in La Reunion belongs to the ESCA genotype (Pialoux et al. 2007; Arankalle et al. 2007), while the strains currently circulating in Southeast Asia include both Asian and ESCA genotypes (Lanciotti and Valadere 2014; Sy et al. 2016; Weaver and Forrester 2015; Powers et al. 2000; Schuffenecker et al. 2006). The strain introduced into Saint Martin and the Americas belongs to the Asian genotype (Leparc-Goffart et al. 2014; Lanciotti and Valadere 2014; Sy et al. 2016), while some evidence suggest that viruses isolated in Brazil in 2014 have an ESCA genotype resembling strains circulating in Angola (Nunes et al. 2015).

Infection with CHIKV causes Chikungunya fever (CHIKF) and symptoms displayed by febrile patients are difficult to distinguish from other arthropod-borne viruses (arboviruses) such as other arthralgic alphaviruses (notably Ross River, Semliki Forest, o'nyong nyong, and Sindbis viruses) and some flaviviruses (dengue and Zika viruses)(Atkins 2013). Patients generally present with severe fever, rash, and joint pain during the acute phase that are usually self-resolving in one to two weeks (Miner et al. 2015). Typical symptoms of CHIKF also include swollen joints (Atkins 2013) and 10–60% of cases lead to persistent arthralgia that could persist weeks, month, or even years (Borgherini et al. 2008; Gerardin et al. 2011; Schilte et al. 2013).

Mortality associated with CHIKF remains low with less than 1 in 1000 cases (Josseran et al. 2006), but complications could arise with observed cases of encephalitis, encephalopathy, myocarditis, hepatitis and multi-organ failures (Wielanek et al. 2007; Schilte et al. 2013; Rajapakse et al. 2010; Lemant et al. 2008). Mortality and morbidity have also been associated with co-morbidities such as age or immune competence (Hoarau et al. 2010).

Despite the on-going efforts, available treatments and vaccines remain limited (Weaver et al. 2012). The development of potent and safe antivirals for prophylaxis, treatment and transmission prevention will need crucial advances in understanding the complex mechanisms of hosts–alphavirus interactions. On one hand, successful development of treatment strategies for patients will involve a deeper understanding of key immune processes involved in driving pathology in patients. On the other hand, successful strategies to prevent new epidemics and impede local transmission will involve a better understanding of vector biology and vector-virus interplays with invertebrate immunity.

In this review, available CHIKV research tools are illustrated with various examples on how useful they have been in the study of virus–mammal and virus–vector interactions that led to a better understanding of CHIKV transmission and pathogenesis.

2 Chikungunya Virus Molecular Tools

CHIKV is a positive-strand RNA virus with a genome of approximately 11.8 kb, flanked by a capped 5' untranslated region (UTR) and a 3' UTR followed by a poly-A tail (Weaver et al. 2012) (Fig. 1a). The genome comprises of two open reading frames coding for two polyproteins: the first is translated into non-structural proteins (nsP) directly from the genomic RNA, while the second codes for the structural proteins and is translated from the 26S subgenomic RNA produced during viral RNA replication by the non-structural proteins (Kuhn 2007).

Since the first reported Sindbis virus infectious clone in 1987 (Rice et al. 1987), alphavirus infectious clones have been significantly improved, but the general principle remains. Typically, the viral genome of the specific virus strain is reverse transcribed and inserted into a DNA plasmid. There are then two possibilities to produce viral particles: (i) DNA transfection in cells if the viral genome is flanked by a promoter (i.e., CMV as represented in Fig. 1a) and a transcription terminator after the poly-A tail (i.e., HDV ribozyme) (Steel et al. 2011; Tretyakova et al. 2014).

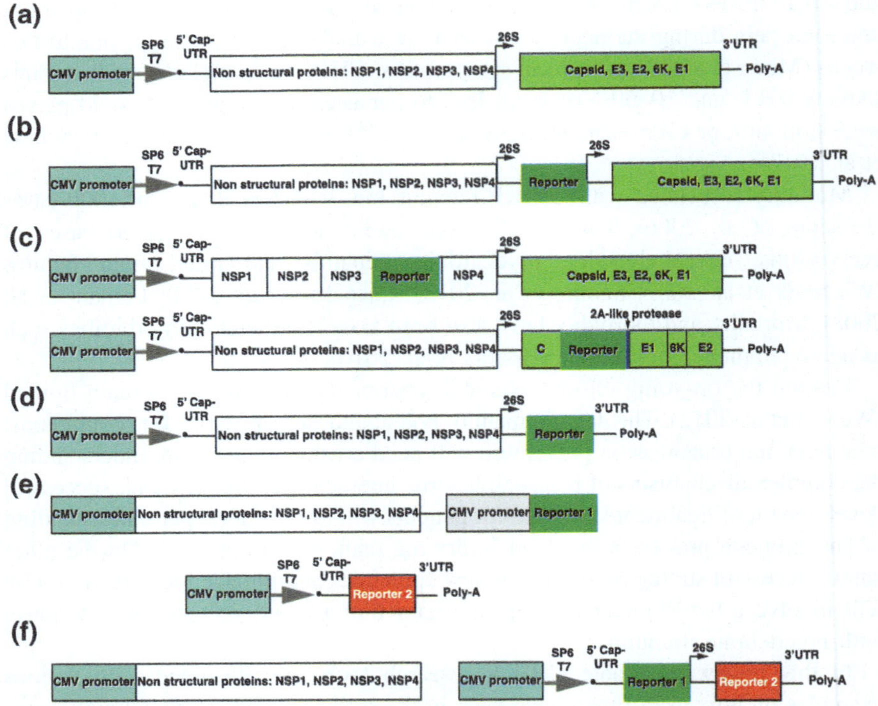

Fig. 1 CHIKV infectious clones and replicons. **a** Infectious clone; **b** infectious clone expressing a reporter gene in a duplicated subgenomic promoter; **c** infectious clone expressing a reporter gene in in-frame of a non-structural (top) or structural (bottom) protein; **d** non-infectious replicon expressing a reporter gene in place of structural proteins; **e–f** trans-replicon systems allowing uncoupling of non-structural proteins from viral RNA replication

(ii) Transfection of in vitro transcribed RNA from a T7 or sp6 promoter (Fig. 1a) from the linearized DNA template (Rice et al. 1987).

The genome organization of alphaviruses makes them relatively tolerant to genetic manipulation. This particularity allows for duplication of the 26S promoter region, allowing for heterologous genes to be expressed from the virus in the same fashion as the viral structural proteins (Fig. 1b) (Frolov et al. 1997). This has been used to express reporter genes as well as cellular proteins or antisense RNAs (Olson et al. 1994; Pushko et al. 2001; Cook and Griffin 2003; Pierro et al. 2003; Travanty et al. 2004; Cirimotich et al. 2009; Phillips et al. 2010). The nature of the reporter gene can be of several types and depends on the readout requirements of the experimental setup. Using CKIHV as an example, a fluorescent reporter gene could be used to discriminate infected and non-infected cells during in vitro experiments such as a genome-wide screen (Karlas et al. 2016), as well as during in vivo experiments using, for example, immunohistochemistry to identify permissive infected cell populations in mouse models (Schilte et al. 2010). Similarly, a

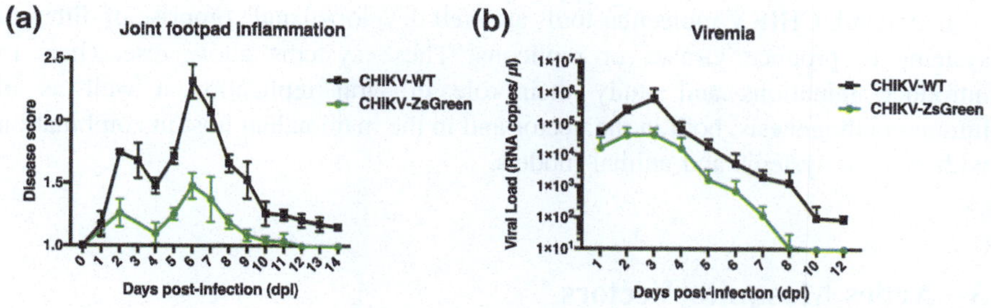

Fig. 2 Comparison of wild type CHIKV isolate (WT) and a CHIKV infectious clone containing ZsGreen (ZsGreen). Four-weeks-old C57BL/6 female mice were infected via ventral footpad inoculation with 10^6 pfu of CHIKV either WT or with ZsGreen downstream of a duplicated subgenomic promoter (Fig. 1b). **a** Disease score of infected footpads and **b** tail vein blood viremia assessed by RT-qPCR. (Carissimo unpublished)

luciferase gene could be used to easily assess virus infection during in vitro antiviral assays using luciferase (Varghese et al. 2016), and to quantify and localize CHIKV infection and replication in animal models (Teo et al. 2013). However, it is important to note that viruses with a duplicated subgenomic promoter would display a slightly delayed replication rate coupled with reduced pathology when compared to its non-modified counterpart (Fig. 2). One way to reduce the issue of a duplicated subgenomic promoter is through the use of an infectious clone with the reporter gene expressed in-frame of a non-structural gene (typically nsP3) or a structural gene (typically caspid) (Fig. 1c) (Sun et al. 2014).

Alternative approaches to reporter infectious clones include the replicon system (Fig. 1d), which has the advantage of lowering safety requirements since no infectious particles are produced (Rayner et al. 2002). Furthermore, the replicon system also allows the study of single-round infections with a system of virus-like replicon particles (Tsetsarkin and Weaver 2011). This process involves packaging pseudo-viral particles containing a replicon genome using cells constitutively expressing viral structural proteins, or by co-transfecting structural protein expression plasmids with the replicon.

The CHIKV replicon system has been further improved and exploited to study specific steps in viral RNA replication (Hellstrom et al. 2017; Utt et al. 2016). One variation involves the uncoupling of the non-structural protein synthesis from RNA replication to allow for easy normalization of viral RNA replication and translation (Fig. 1e). Alternatively, a modified replicon with a template containing one reporter gene for nsPs and another reporter gene for structural proteins will allow the effect of nsP mutations on the relative expression levels of genomic (non-structural ORF) and subgenomic (structural ORF) RNA to be studied (Fig. 1f). Mutations affecting viral replication can induce important negative feedback loops, which this uncoupling can address. Such mutations induce compensatory mutations or revertants in infectious clones when nsPs are expressed from viral RNA (Lulla et al. 2012, 2013; Utt et al. 2016).

In general, CHIKV molecular tools are well developed and comprise of different systems to produce viruses or replicons. These systems allow researchers to introduce mutations and study their role in viral replication a well as in immunopathogenesis, both in the vector and in the mammalian host in combination with in vitro systems and animal models.

3 Aedes Mosquito Vectors

CHIKV is transmitted mainly by a previously infected female mosquito of the *Aedes* sp. during blood feeding on a mammalian host. In human transmission cycles, *Aedes aegypti* and *Aedes albopictus* are the main vectors (Dubrulle et al. 2009; Vazeille et al. 2007). Understanding the transmission dynamics is an important parameter to control epidemics and future outbreaks since the current primary control strategy for arboviruses is mosquito population control (Diallo et al. 2016).

3.1 Aedes Mosquito In Vitro Cell Line Systems

In contrast to mammalian hosts, there are no extensive and well-characterized in vitro systems for mosquito vectors (Walker et al. 2014). Cell lines from *Aedes* (*aegypti* and *albopictus*) derived mainly from larval stages are typically not completely homogenous (Walker et al. 2014). Thus, the heterogeneity of these mosquito cell lines could sometimes lead to a lot of uncertainties in the observed anti- or pro-viral effects due to the different cell types involved. The C6/36 clonal line (ATCC CRL-1660) was initially selected for its uniformity in producing high viral yield from a progenitor cell line established from freshly hatched *aedes albopictus* larvae of unspecified ancestry (Singh and Pavri 1967; Igarashi 1978). This cell line is probably the most widely used for virus production since it allows high viral titer productions due to a defective antiviral RNA interference (RNAi) pathway (Brackney et al. 2010). Recent genome analysis has reported it to be of male origin (Miller et al. 2017), further questioning its relevance in studying female mosquito antiviral immunity.

Nevertheless, using these cell lines, studies have characterized important components of the antiviral response in mosquitoes. RNAi has been demonstrated in various models to be an important part of the antiviral response in *Aedes* mosquitoes (Blair 2011). Consistent with these results, CHIKV nsP2 and nsP3 were recently found to possess RNAi suppressor activities (Mathur et al. 2016). Further studies would be needed to delineate the different small RNA pathways involved and their roles in vector antiviral defense. More recently, a study on an orthobunyavirus showed that different components of the RNAi pathways were selectively involved in the antiviral defense against different viruses, suggesting a specialization of small RNA pathways in the different RNAi antiviral defense mechanisms (Dietrich et al. 2017).

Surprisingly, other pathways of the innate immunity in invertebrates, Toll, Imd, and JAK/STAT pathways have not been demonstrated to possess in vitro antiviral activities against CHIKV (McFarlane et al. 2014). However, CHIKV was shown to possess the ability to down-regulate the Toll pathway, suggesting that these pathways could play a role in vivo in the vector antiviral defense (McFarlane et al. 2014).

3.2 *In Vivo Studies in Mosquito Vectors*

Mosquito vectorial capacity is an important part of arboviral transmission dynamics (Hardy et al. 1983), and depends on population density, blood feeding behavior, vector longevity, and vector competence. Vector competence is defined as the ability of an arthropod vector midgut to get infected by the virus through a blood meal, and establish a persistent infection that can disseminate and reach the salivary glands to get transmitted via saliva during the subsequent blood meal (Hardy et al. 1983; Kramer and Ciota 2015; Lounibos and Kramer 2016). Studies performed in *Anopheles* mosquitoes, a vector of the alphavirus o'nyong nyong virus (ONNV), phylogenetically closest relative to CHIKV (Powers et al. 2001), showed a high compartmentalization of the antiviral immune response and allowed dissection of the infection kinetics and organ tropism of this virus (Carissimo et al. 2015). This study showed an important difference in the involvement of the mosquito RNAi pathway as an antiviral pathway depending on the viral administration route and the infected organ. Indeed, previous studies using needle infection of the vector had occulted the fundamental differences between the midgut and system immunity by bypassing the first immune compartment (Keene et al. 2004; Waldock et al. 2012). This example highlights the importance of studies involving the natural infection route in animal models and the importance of studying immune response in the correct cell types at the correct time points.

Research in mosquito vectors has generated interesting knowledge on the role of the invertebrate RNAi system (Olson and Blair 2015; Balakrishna Pillai et al. 2017). RNAi is described as the major antiviral pathway that controls arbovirus infections in vectors (Olson and Blair 2015; Balakrishna Pillai et al. 2017). It is responsible for controlling viral load against CHIKV, SINV, and DENV in *Aedes* mosquitoes (McFarlane et al. 2014; Khoo et al. 2010, 2013). Interestingly, a study of the genetic diversity of Dicer 2, a key RNAi gene showed that Dicer 2 genetic sequences have an evolutionary signature correlating with specific mosquito susceptibilities to DENV isolates (Bernhardt et al. 2012; Lambrechts et al. 2013). This discovery is a representative of the genotype-by-genotype interactions that can be studied in vector–arbovirus interactions.

In addition to its role as an important immune effector in the antiviral response in mosquitoes, RNAi is also a powerful tool that allows for in vivo functional genomic studies by simply injecting double-stranded RNA targeting genes of interest. Surprisingly, this powerful approach has not been used extensively to study in vivo *Aedes* immune system interactions with CHIKV. However, it has been used successfully to elucidate the role of several immune pathways against other arboviruses: (i) Toll immune pathway has been shown to control DENV via silencing of Cactus, a negative regulator of the pathway, or the silencing of MyD88, an upstream adapter protein (Xi et al. 2008). The JAK/STAT pathway activated during *Aedes* midgut infection was shown to regulate DENV infection via Hop and Dome activation, while downregulation of the negative JAK/STAT regulator PIAS inhibited virus infection (Souza-Neto et al. 2009).

CHIKV infectious clones have also allowed researchers to perform point mutations in the viral genome and study their consequences on the vectorial capacity in *Aedes* mosquitoes (Tsetsarkin et al. 2009, 2007; Tsetsarkin and Weaver 2011). Notably, the A226 V mutation introduced back into an infectious clone of the CHIKV strain isolated before the La Reunion epidemic in 2006, provided instrumental knowledge in the confirmatory role of the epidemic strain in vector adaptation to the *Aedes albopictus* (Tsetsarkin et al. 2007). Importantly, sequential mutations with an epistatic effect were subsequently found to be crucial for vector adaptation (Tsetsarkin et al. 2009). Specifically, the role of amino acid substitutions in the envelope proteins E1 and E2 were assessed on the midgut infection barrier, using single-round infections with virus-like particles (VLPs) containing replicon-expressing reporter genes. This novel approach allowed the demonstration that these mutations increased initial infection of *Aedes albopictus* midgut cells during blood meal infection (Tsetsarkin and Weaver 2011). Recently, co-infection studies using CHIKV and ZIKV or CHIKV and DENV showed that simultaneous exposure to viruses in blood meals did not modify infection and transmission rates of these viruses by the vectors (Goertz et al. 2017; Le Coupanec et al. 2017). Interestingly, it was also highlighted that *Aedes* mosquitoes had a strong salivary gland barrier to CHIKV and a less profound barrier for ZIKV (Goertz et al. 2017).

4 Mammalian Hosts

CHIKF has been reported to be asymptomatic in 15% of cases (Miner et al. 2015). Symptoms typically start with acute onset of high fever (>39 °C) accompanied by myalgia, arthralgia, and in some cases a maculopapular rash with a wide range of severity. Between 32 and 95% of patients are present with pain and swelling of the joints (>1), usually symmetrical, during the acute phase (Burt et al. 2017). However, in a significant number of patients, debilitating inflammatory arthritis symptoms could persist after the acute phase and have been reported to last up to 3 years after infection (Marimoutou et al. 2012; Schilte et al. 2013). The health and socio-economical impact of these persisting debilitating symptoms on exposed populations warrants research that identifies the causes and offers therapeutic avenues to explore.

4.1 Viral Proteins Functions and Mechanisms of Replication

For mechanistic studies in the mammalian host, cell cultures have been essential. They have helped elucidate some of the multiple functions of the different viral proteins and the roles and mechanisms of antiviral-host proteins. Overexpression of viral proteins or the use of mutated, or chimera, viruses has been the choice strategy.

Non-structural proteins are translated from the genomic viral RNA as a nsP1234 polyprotein precursor subsequently cleaved by the protease activity of nsP2. In vitro studies have shown that nsP1 binds to membranes and possesses methyl- and guanylyltransferase activities required for viral RNA capping (Ahola and Karlin 2015). nsP1 is necessary to counteract the antiviral effect of Tetherin at the plasma membrane (Jones et al. 2013). In addition to the direct NTPase, helicase, and protease activities on viral replication, nsP2 is also highly connected to host proteins and allows the inhibition of host cell mRNA transcription, JAK/STAT signaling, unfolded protein response, and autophagy (Karpe et al. 2011; Das et al. 2014; Utt et al. 2015; Bourai et al. 2012; Akhrymuk et al. 2012; Fros et al. 2010, 2013, 2015b; Judith et al. 2013). nsP3 is an important viral protein that can bind RNA and interacts with several important host factors including G3BP proteins essential for viral infectivity (Malet et al. 2009; Panas et al. 2014). Remarkably, nsP3 was also demonstrated to interact with mosquito G3BP homolog Rasputin and to be essential for in vivo infectivity (Fros et al. 2015a). nsP4 is hypothesized to be the RNA-dependent RNA polymerase of CHIKV by homology of this activity in other alphaviruses and has been shown to be able to suppress host antiviral pathways (Rathore et al. 2013).

Replicon and trans-replicon systems have successfully been used to understand the assembly of the viral replication complex (membrane invaginations formed by the non-structural proteins and viral RNA where viral RNA replication occurs) and the function of the regulation of nsP2 cleavage of the nsP1234 precursors in this assembly of replication complexes (Hellstrom et al. 2017; Utt et al. 2016). These studies pave the way for the design of intelligent antiviral screens designed to inhibit specific viral protein functions essential for viral proteins functions and viral replication complex morphology, or to inhibit viral–host proteins interaction essential for viral replication.

4.2 Genome-Wide Screens and Targeting the Mammalian Host

Genome-wide RNAi screens have been performed in mammalian systems to identify host factors with pro- or antiviral effects on CHIKV (Karlas et al. 2016; Panda et al. 2013; Ooi et al. 2013). These screens have yielded interesting results in identification of potential druggable host targets with the identification of two pathway inhibitors having prophylactic antiviral effects on CHIKV (Karlas et al. 2016). In that study, authors validated their genome-wide screening results by showing that the prophylactic effects against CHIKV were reproducible in mouse models.

However, these screens have pitfalls: pro- and anti-viral host factors identified can vary depending on the experimental design, the viral strain (or infectious clone) used, the cells used for infection, the detection read-out, and known gene coding. In addition, these assays exclude all host factors that impair cell viability, which precludes unbiased identification of host factors essential for virus and host cell biology.

4.3 Human Patient Cohorts and Immune Mediators

In addition to retrospective studies and case studies, patient cohorts have been used to identify the immune mediators regulated during viral infection in patient sera. These studies use ELISA and multiplex ELISA techniques such as Luminex and bead-based arrays to quantify specific analytes at different time points and correlate them to patient symptom severities and outcomes. The different techniques and variety of commercial reagents make these studies difficult to compare. Nevertheless, a recent meta-analysis of 14 different patient cohorts allowed identification of a common signature of immune mediators in infected patients (Teng et al. 2015). Identification of a specific pro-inflammatory cytokine signature is a first step in understanding by other methods the immune mechanisms leading to the debilitating symptoms. For example, patient immune mediator quantification during acute CHIKV infection showed increased levels of IL-18 and IL-1b, known downstream indicators of inflammasome activation (Chen et al. 2017). This study led to the realization that inflammasome inhibitors abrogated CHIKV and Ross river virus-induced joint swelling, bone loss, and myositis in mouse models, opening a potential treatment avenue for symptoms associated with arthralgic alphavirus infection in patients (Chen et al. 2017).

However, soluble immune mediators detectable in patient serum or plasma are a limited source of information to understand the underlying pathogenic immune mechanism at play in the joints of human patients, especially the potential viral persistence and the mechanisms at play in the chronic debilitating phase of the disease.

4.4 Protective Immunity in Animal Models

In order to circumvent limitations in patient cohorts, animal models have been used to decipher the immune cells and immune processes involved in CHIKV pathology.

Two main animal models exist to study CHIKV immunopathogenesis: non-human primate (NHP) models and mouse models (Broeckel et al. 2015; Haese et al. 2016). Both systems have advantages and pitfalls.

Mouse models, while they do not reproduce all human symptoms, allow for mechanistic studies on the role of specific cell types and specific genes in virus-induced disease. Unfortunately, until now, no mouse model mimics the inflammatory arthritis observed in patients. The closest model would be in C56BL/6 althralgia mouse models where viral persistence in the joints of virus-infected animals has been observed (Poo et al. 2014b; Teo et al. 2013). Adult mice deficient for adaptive immunity (B- and T-cell deficient) do not succumb to infection despite high viral burden in blood and organs (Hawman et al. 2013). CHIKV-specific antibodies were shown to be present rapidly after infection and to participate in viremia clearance (Lum et al. 2013). CD4 T-cells were shown to be pathogenic in the joint while being involved in the quantity and quality of the anti-CHIKV neutralizing antibodies (Lum et al. 2013; Teo et al. 2013). CCR2 monocytes and macrophages have a protective and pathogenic role in the joint pathology severity (Chen et al. 2015; Poo et al. 2014a). A role for other cell types has also been suggested (Poo et al. 2014b). Expanding the regulatory T-cell repertoire to control CD4 T-cell pathogenicity or reduce their migration in the joint can reduce the joint pathology, and this could be an avenue for future treatment (Lee et al. 2015; Teo et al. 2017).

Long-term viral persistence in the joints is one of the hypotheses for the persisting arthritis observed in patients, and a deeper understanding of the mechanism of adaptive immunity leading to this phenomenon can lead to novel therapeutic approaches.

4.5 Innate Immunity in Animal Models

Studies in mouse models have also highlighted the importance of innate immunity and the role of specific genes in CHIKV infection. Mice deficient for functional type I IFN pathway are highly susceptible to CHIKV infection (Couderc et al. 2008; Schilte et al. 2010; Gardner et al. 2012; Rudd et al. 2012; Schilte et al. 2012). In vitro experiments and mouse models have allowed the understanding of pathogenic or protective roles of specific genes in the anti-CHIKV immune response. The antiviral role of interferon-stimulated genes such as ISG15, Tetherin, or Viperin against CHIKV was shown (Werneke et al. 2011; Mahauad-Fernandez et al. 2014; Teng et al. 2012), as well as the role of specific pattern recognition receptors such as TLR3 or RIG-I (Her et al. 2015; Priya et al. 2014; Olagnier et al. 2014). Using a neonate model and bone marrow chimeras, Schilte and colleagues showed the role of IRF3 and IRF7 in non-hematopoietic cells in anti-CHIKV immune responses (Schilte et al. 2012). However, deeper understanding of the role of these genes in shaping the innate and adaptive immunity against CHIKV is needed and will allow better treatment and vaccine designs.

4.6 Development and Assessment of Therapeutics in Animal Models

Mouse models are useful to dissect the role of specific genes or specific cellular response to CHIKV infection, but do not recapitulate all the human symptoms. However, NHPs, in particular cynomolgus macaques, recapitulate very similar features (viral, clinical, and pathological) to human patients when infected with CHIKV (Labadie et al. 2010). Persistence of virus antigen was also observed in various tissues and lymphoid organs up to 3 months post infection (Labadie et al. 2010), suggesting that viral persistence in tissues could be responsible for the chronic symptoms in human patients. Pregnant macaques were used to study a potential vertical transmission of CHIKV (Chen et al. 2010), and showed that human cases of vertical transmission were likely the result of contamination during delivery rather than in utero. Macaques were also used to evaluate age-related immunity in CHIKV infection where immune senescence was demonstrated to influence anti-CHIKV response (Messaoudi et al. 2013). All these studies clearly illustrated that macaque models could recapitulate human symptoms of CHIKF (Chen et al. 2010; Labadie et al. 2010; Messaoudi et al. 2013). However, due to costs concerns, these models have been used mostly for evaluation of preclinical studies of vaccines or therapeutics at a more developed stage.

5 Conclusions

The non-exhaustive tools presented in this review have allowed researchers to progress immensely in the comprehension of the mechanisms surrounding CHIKV infections Vector biology studies, vector competence assessments, and studies focused on interaction of CHIKV with invertebrate pro- and antiviral factors and immunity of *Aedes sp.*, have contributed to the comprehension of epidemic dynamics. In vitro studies, animal models, and human patient cohorts have advanced the comprehension of the interplay between innate and adaptive immunity leading to CHIKV pathology. The role of different interferon-stimulated genes and the identification of pro- and antiviral mechanisms in humans were also defined. To date, these studies have not yielded any licensed antiviral drug or vaccine against CHIKV, but promising results have been obtained in animal models. Current treatment options during CHIKV infection consist of anti-inflammatory compounds for relief of debilitating symptoms. In mouse models, antivirals and immune pathway inhibitors are promising, but target different phases of the disease should be better evaluated. Some viral replication inhibitors, inflammatory arthritis inhibitors, and vaccine candidates are in preclinical phase.

Arboviruses, having a viral life cycle alternating between mammalian host and mosquito vector, are well-tuned pathogens specialized in the subversion of cellular processes conserved in both hosts, with a narrow evolutionary space to conserve efficient replication and transmission. Identification of these conserved viral–host factors is important since they would be very interesting and promising candidates in the development of dual role therapeutic strategies, allowing an inhibition of viral replication in patients as well as transmission reducing agents with a reduced possibility of viral escape mechanisms. Therefore, the development of high-throughput unbiased tools to study relevant host factors or pathways in both humans and mosquito vectors will be essential for implementing strategies to develop small inhibitors, active against CHIKV infections in both hosts.

References

Ahola T, Karlin DG (2015) Sequence analysis reveals a conserved extension in the capping enzyme of the alphavirus supergroup, and a homologous domain in nodaviruses. Biol Direct 10:16. https://doi.org/10.1186/s13062-015-0050-0

Akhrymuk I, Kulemzin SV, Frolova EI (2012) Evasion of the innate immune response: the Old World alphavirus nsP2 protein induces rapid degradation of Rpb1, a catalytic subunit of RNA polymerase II. J Virol 86(13):7180–7191. https://doi.org/10.1128/JVI.00541-12

Arankalle VA, Shrivastava S, Cherian S, Gunjikar RS, Walimbe AM, Jadhav SM, Sudeep AB, Mishra AC (2007) Genetic divergence of Chikungunya viruses in India (1963–2006) with special reference to the 2005–2006 explosive epidemic. J Gen Virol 88(Pt 7):1967–1976. https://doi.org/10.1099/vir.0.82714-0

Atkins GJ (2013) The pathogenesis of alphaviruses. ISRN. Virology 2013:22. https://doi.org/10.5402/2013/861912

Balakrishna Pillai A, Nagarajan U, Mitra A, Krishnan U, Rajendran S, Hoti SL, Mishra RK (2017) RNA interference in mosquito: understanding immune responses, double-stranded RNA delivery systems and potential applications in vector control. Insect Mol Biol 26(2):127–139. https://doi.org/10.1111/imb.12282

Bernhardt SA, Simmons MP, Olson KE, Beaty BJ, Blair CD, Black WC (2012) Rapid intraspecific evolution of miRNA and siRNA genes in the mosquito *Aedes aegypti*. PLoS ONE 7(9): e44198. https://doi.org/10.1371/journal.pone.0044198

Blair CD (2011) Mosquito RNAi is the major innate immune pathway controlling arbovirus infection and transmission. Future Microbiol 6(3):265–277. https://doi.org/10.2217/fmb.11.11

Borgherini G, Poubeau P, Jossaume A, Gouix A, Cotte L, Michault A, Arvin-Berod C, Paganin F (2008) Persistent arthralgia associated with Chikungunya virus: a study of 88 adult patients on reunion island. Clin Infect Dis 47(4):469–475. https://doi.org/10.1086/590003

Bourai M, Lucas-Hourani M, Gad HH, Drosten C, Jacob Y, Tafforeau L, Cassonnet P, Jones LM, Judith D, Couderc T, Lecuit M, Andre P, Kummerer BM, Lotteau V, Despres P, Tangy F, Vidalain PO (2012) Mapping of Chikungunya virus interactions with host proteins identified nsP2 as a highly connected viral component. J Virol 86(6):3121–3134. https://doi.org/10.1128/JVI.06390-11

Brackney DE, Scott JC, Sagawa F, Woodward JE, Miller NA, Schilkey FD, Mudge J, Wilusz J, Olson KE, Blair CD, Ebel GD (2010) C6/36 Aedes albopictus cells have a dysfunctional antiviral RNA interference response. PLoS Negl Trop Dis 4(10):e856. https://doi.org/10.1371/journal.pntd.0000856

Broeckel R, Haese N, Messaoudi I, Streblow DN (2015) Nonhuman primate models of Chikungunya virus infection and disease (CHIKV NHP Model). Pathogens 4(3):662–681. https://doi.org/10.3390/pathogens4030662

Burt FJ, Chen W, Miner JJ, Lenschow DJ, Merits A, Schnettler E, Kohl A, Rudd PA, Taylor A, Herrero LJ, Zaid A, Ng LFP, Mahalingam S (2017) Chikungunya virus: an update on the

biology and pathogenesis of this emerging pathogen. Lancet Infect Dis 17(4):e107–e117. https://doi.org/10.1016/S1473-3099(16)30385-1
Carey DE (1971) Chikungunya and dengue: a case of mistaken identity? J Hist Med Allied Sci 26 (3):243–262
Carissimo G, Pondeville E, McFarlane M, Dietrich I, Mitri C, Bischoff E, Antoniewski C, Bourgouin C, Failloux AB, Kohl A, Vernick KD (2015) Antiviral immunity of Anopheles gambiae is highly compartmentalized, with distinct roles for RNA interference and gut microbiota. Proc Natl Acad Sci U S A 112(2):E176–E185. https://doi.org/10.1073/pnas.1412984112
Chen CI, Clark DC, Pesavento P, Lerche NW, Luciw PA, Reisen WK, Brault AC (2010) Comparative pathogenesis of epidemic and enzootic Chikungunya viruses in a pregnant Rhesus macaque model. Am J Trop Med Hyg 83(6):1249–1258. https://doi.org/10.4269/ajtmh.2010.10-0290
Chen W, Foo SS, Taylor A, Lulla A, Merits A, Hueston L, Forwood MR, Walsh NC, Sims NA, Herrero LJ, Mahalingam S (2015) Bindarit, an inhibitor of monocyte chemotactic protein synthesis, protects against bone loss induced by Chikungunya virus infection. J Virol 89 (1):581–593. https://doi.org/10.1128/JVI.02034-14
Chen W, Foo SS, Zaid A, Teng TS, Herrero LJ, Wolf S, Tharmarajah K, Vu LD, van Vreden C, Taylor A, Freitas JR, Li RW, Woodruff TM, Gordon R, Ojcius DM, Nakaya HI, Kanneganti TD, O'Neill LAJ, Robertson AAB, King NJ, Suhrbier A, Cooper MA, Ng LFP, Mahalingam S (2017) Specific inhibition of NLRP3 in Chikungunya disease reveals a role for inflammasomes in alphavirus-induced inflammation. Nat Microbiol. https://doi.org/10.1038/s41564-017-0015-4
Cirimotich CM, Scott JC, Phillips AT, Geiss BJ, Olson KE (2009) Suppression of RNA interference increases alphavirus replication and virus-associated mortality in *Aedes aegypti* mosquitoes. BMC Microbiol 9:49. https://doi.org/10.1186/1471-2180-9-49
Cook SH, Griffin DE (2003) Luciferase imaging of a neurotropic viral infection in intact animals. J Virol 77 (9):5333–5338
Cordel H, Quatresous I, Paquet C, Couturier E (2006) Imported cases of Chikungunya in metropolitan France, April 2005–February 2006. Euro Surveill 11 (4):E060420 060423
Couderc T, Chretien F, Schilte C, Disson O, Brigitte M, Guivel-Benhassine F, Touret Y, Barau G, Cayet N, Schuffenecker I, Despres P, Arenzana-Seisdedos F, Michault A, Albert ML, Lecuit M (2008) A mouse model for Chikungunya: young age and inefficient type-I interferon signaling are risk factors for severe disease. PLoS Pathog 4(2):e29. https://doi.org/10.1371/journal.ppat.0040029
Das PK, Merits A, Lulla A (2014) Functional cross-talk between distant domains of Chikungunya virus non-structural protein 2 is decisive for its RNA-modulating activity. J Biol Chem 289 (9):5635–5653. https://doi.org/10.1074/jbc.M113.503433
Diallo M, Dia I, Diallo D, Diagne CT, Ba Y, Yactayo S (2016) Perspectives and challenges in entomological risk assessment and vector control of Chikungunya. J Infect Dis 214(suppl 5): S459–S465. https://doi.org/10.1093/infdis/jiw397
Dietrich I, Shi X, McFarlane M, Watson M, Blomstrom AL, Skelton JK, Kohl A, Elliott RM, Schnettler E (2017) The antiviral RNAi response in vector and non-vector cells against Orthobunyaviruses. PLoS Negl Trop Dis 11(1):e0005272. https://doi.org/10.1371/journal.pntd.0005272
Dubrulle M, Mousson L, Moutailler S, Vazeille M, Failloux AB (2009) Chikungunya virus and Aedes mosquitoes: saliva is infectious as soon as two days after oral infection. PLoS ONE 4(6): e5895. https://doi.org/10.1371/journal.pone.0005895
Frolov I, Frolova E, Schlesinger S (1997) Sindbis virus replicons and Sindbis virus: assembly of chimeras and of particles deficient in virus RNA. J Virol 71(4):2819–2829
Fros JJ, Geertsema C, Zouache K, Baggen J, Domeradzka N, van Leeuwen DM, Flipse J, Vlak JM, Failloux AB, Pijlman GP (2015a) Mosquito Rasputin interacts with Chikungunya virus nsP3 and determines the infection rate in *Aedes albopictus*. Parasit Vectors 8:464. https://doi.org/10.1186/s13071-015-1070-4

Fros JJ, Liu WJ, Prow NA, Geertsema C, Ligtenberg M, Vanlandingham DL, Schnettler E, Vlak JM, Suhrbier A, Khromykh AA, Pijlman GP (2010) Chikungunya virus nonstructural protein 2 inhibits type I/II interferon-stimulated JAK-STAT signaling. J Virol 84(20):10877–10887. https://doi.org/10.1128/JVI.00949-10

Fros JJ, Major LD, Scholte FE, Gardner J, van Hemert MJ, Suhrbier A, Pijlman GP (2015b) Chikungunya virus non-structural protein 2-mediated host shut-off disables the unfolded protein response. J Gen Virol 96(Pt 3):580–589. https://doi.org/10.1099/vir.0.071845-0

Fros JJ, van der Maten E, Vlak JM, Pijlman GP (2013) The C-terminal domain of Chikungunya virus nsP2 independently governs viral RNA replication, cytopathicity, and inhibition of interferon signaling. J Virol 87(18):10394–10400. https://doi.org/10.1128/JVI.00884-13

Gardner CL, Burke CW, Higgs ST, Klimstra WB, Ryman KD (2012) Interferon-alpha/beta deficiency greatly exacerbates arthritogenic disease in mice infected with wild-type Chikungunya virus but not with the cell culture-adapted live-attenuated 181/25 vaccine candidate. Virology 425(2):103–112. https://doi.org/10.1016/j.virol.2011.12.020

Gerardin P, Fianu A, Malvy D, Mussard C, Boussaid K, Rollot O, Michault A, Gauzere BA, Breart G, Favier F (2011) Perceived morbidity and community burden after a Chikungunya outbreak: the TELECHIK survey, a population-based cohort study. BMC Med 9:5. https://doi.org/10.1186/1741-7015-9-5

Goertz GP, Vogels CBF, Geertsema C, Koenraadt CJM, Pijlman GP (2017) Mosquito co-infection with Zika and Chikungunya virus allows simultaneous transmission without affecting vector competence of *Aedes aegypti*. PLoS Negl Trop Dis 11(6):e0005654. https://doi.org/10.1371/journal.pntd.0005654

Haese NN, Broeckel RM, Hawman DW, Heise MT, Morrison TE, Streblow DN (2016) Animal models of Chikungunya virus infection and disease. J Infect Dis 214(suppl 5):S482–S487. https://doi.org/10.1093/infdis/jiw284

Halstead SB (2015) Reappearance of chikungunya, formerly called dengue, in the Americas. Emerg Infect Dis 21(4):557–561. https://doi.org/10.3201/eid2104.141723

Hapuarachchi HC, Bandara KB, Sumanadasa SD, Hapugoda MD, Lai YL, Lee KS, Tan LK, Lin RT, Ng LF, Bucht G, Abeyewickreme W, Ng LC (2010) Re-emergence of Chikungunya virus in South-east Asia: virological evidence from Sri Lanka and Singapore. J Gen Virol 91(Pt 4):1067–1076. https://doi.org/10.1099/vir.0.015743-0

Hardy JL, Houk EJ, Kramer LD, Reeves WC (1983) Intrinsic factors affecting vector competence of mosquitoes for arboviruses. Annu Rev Entomol 28:229–262. https://doi.org/10.1146/annurev.en.28.010183.001305

Hawman DW, Stoermer KA, Montgomery SA, Pal P, Oko L, Diamond MS, Morrison TE (2013) Chronic joint disease caused by persistent Chikungunya virus infection is controlled by the adaptive immune response. J Virol 87(24):13878–13888. https://doi.org/10.1128/JVI.02666-13

Hellstrom K, Kallio K, Utt A, Quirin T, Jokitalo E, Merits A, Ahola T (2017) Partially uncleaved alphavirus replicase forms spherule structures in the presence and absence of RNA template. J Virol 91(18). https://doi.org/10.1128/jvi.00787-17

Her Z, Teng TS, Tan JJ, Teo TH, Kam YW, Lum FM, Lee WW, Gabriel C, Melchiotti R, Andiappan AK, Lulla V, Lulla A, Win MK, Chow A, Biswas SK, Leo YS, Lecuit M, Merits A, Renia L, Ng LF (2015) Loss of TLR3 aggravates CHIKV replication and pathology due to an altered virus-specific neutralizing antibody response. EMBO Mol Med 7(1):24–41. https://doi.org/10.15252/emmm.201404459

Hoarau JJ, Jaffar Bandjee MC, Krejbich Trotot P, Das T, Li-Pat-Yuen G, Dassa B, Denizot M, Guichard E, Ribera A, Henni T, Tallet F, Moiton MP, Gauzere BA, Bruniquet S, Jaffar Bandjee Z, Morbidelli P, Martigny G, Jolivet M, Gay F, Grandadam M, Tolou H, Vieillard V, Debre P, Autran B, Gasque P (2010) Persistent chronic inflammation and infection by Chikungunya arthritogenic alphavirus in spite of a robust host immune response. J Immunol 184(10):5914–5927. https://doi.org/10.4049/jimmunol.0900255

Igarashi A (1978) Isolation of a Singh's *Aedes albopictus* cell clone sensitive to Dengue and Chikungunya viruses. J Gen Virol 40(3):531–544. https://doi.org/10.1099/0022-1317-40-3-531

Jones PH, Maric M, Madison MN, Maury W, Roller RJ, Okeoma CM (2013) BST-2/tetherin-mediated restriction of chikungunya (CHIKV) VLP budding is counteracted by CHIKV non-structural protein 1 (nsP1). Virology 438(1):37–49. https://doi.org/10.1016/j.virol.2013.01.010

Josseran L, Paquet C, Zehgnoun A, Caillere N, Le Tertre A, Solet JL, Ledrans M (2006) Chikungunya disease outbreak, Reunion Island. Emerg Infect Dis 12(12):1994–1995. https://doi.org/10.3201/eid1212.060710

Judith D, Mostowy S, Bourai M, Gangneux N, Lelek M, Lucas-Hourani M, Cayet N, Jacob Y, Prevost MC, Pierre P, Tangy F, Zimmer C, Vidalain PO, Couderc T, Lecuit M (2013) Species-specific impact of the autophagy machinery on Chikungunya virus infection. EMBO Rep 14(6):534–544. https://doi.org/10.1038/embor.2013.51

Kalantri SP, Joshi R, Riley LW (2006) Chikungunya epidemic: an Indian perspective. Natl Med J India 19(6):315–322

Karlas A, Berre S, Couderc T, Varjak M, Braun P, Meyer M, Gangneux N, Karo-Astover L, Weege F, Raftery M, Schonrich G, Klemm U, Wurzlbauer A, Bracher F, Merits A, Meyer TF, Lecuit M (2016) A human genome-wide loss-of-function screen identifies effective Chikungunya antiviral drugs. Nat Commun 7:11320. https://doi.org/10.1038/ncomms11320

Karpe YA, Aher PP, Lole KS (2011) NTPase and 5'-RNA triphosphatase activities of Chikungunya virus nsP2 protein. PLoS ONE 6(7):e22336. https://doi.org/10.1371/journal.pone.0022336

Keene KM, Foy BD, Sanchez-Vargas I, Beaty BJ, Blair CD, Olson KE (2004) RNA interference acts as a natural antiviral response to O'nyong-nyong virus (Alphavirus; Togaviridae) infection of Anopheles gambiae. Proc Natl Acad Sci U S A 101(49):17240–17245. https://doi.org/10.1073/pnas.0406983101

Khoo CC, Doty JB, Heersink MS, Olson KE, Franz AW (2013) Transgene-mediated suppression of the RNA interference pathway in *Aedes aegypti* interferes with gene silencing and enhances Sindbis virus and dengue virus type 2 replication. Insect Mol Biol 22(1):104–114. https://doi.org/10.1111/imb.12008

Khoo CC, Piper J, Sanchez-Vargas I, Olson KE, Franz AW (2010) The RNA interference pathway affects midgut infection- and escape barriers for Sindbis virus in *Aedes aegypti*. BMC Microbiol 10:130. https://doi.org/10.1186/1471-2180-10-130

Kramer LD, Ciota AT (2015) Dissecting vectorial capacity for mosquito-borne viruses. Curr Opin Virol 15:112–118. https://doi.org/10.1016/j.coviro.2015.10.003

Kuhn R (2007) Togaviridae: the viruses and their replication. In: Knipe DMHP (ed) Fields' Virology, 5th edn. Lippincott, Williams and Wilkins, NY, USA, pp 1001–1022

Labadie K, Larcher T, Joubert C, Mannioui A, Delache B, Brochard P, Guigand L, Dubreil L, Lebon P, Verrier B, de Lamballerie X, Suhrbier A, Cherel Y, Le Grand R, Roques P (2010) Chikungunya disease in nonhuman primates involves long-term viral persistence in macrophages. J Clin Invest 120(3):894–906. https://doi.org/10.1172/JCI40104

Lambrechts L, Quillery E, Noel V, Richardson JH, Jarman RG, Scott TW, Chevillon C (2013) Specificity of resistance to dengue virus isolates is associated with genotypes of the mosquito antiviral gene Dicer-2. Proc Biol Sci 280(1751):20122437. https://doi.org/10.1098/rspb.2012.2437

Lanciotti RS, Valadere AM (2014) Transcontinental movement of Asian genotype Chikungunya virus. Emerg Infect Dis 20(8):1400–1402. https://doi.org/10.3201/eid2008.140268

Le Coupanec A, Tchankouo-Nguetcheu S, Roux P, Khun H, Huerre M, Morales-Vargas R, Enguehard M, Lavillette D, Misse D, Choumet V (2017) Co-infection of mosquitoes with Chikungunya and dengue viruses reveals modulation of the replication of both viruses in midguts and salivary glands *of Aedes aegypti* mosquitoes. Int J Mol Sci 18(8). https://doi.org/10.3390/ijms18081708

Lee WW, Teo TH, Her Z, Lum FM, Kam YW, Haase D, Renia L, Rotzschke O, Ng LF (2015) Expanding regulatory T cells alleviates Chikungunya virus-induced pathology in mice. J Virol. https://doi.org/10.1128/jvi.00998-15

Lemant J, Boisson V, Winer A, Thibault L, Andre H, Tixier F, Lemercier M, Antok E, Cresta MP, Grivard P, Besnard M, Rollot O, Favier F, Huerre M, Campinos JL, Michault A (2008) Serious acute Chikungunya virus infection requiring intensive care during the Reunion Island outbreak in 2005–2006. Crit Care Med 36(9):2536–2541. https://doi.org/10.1097/CCM.0b013e318183f2d2

Leparc-Goffart I, Nougairede A, Cassadou S, Prat C, de Lamballerie X (2014) Chikungunya in the Americas. Lancet 383(9916):514. https://doi.org/10.1016/S0140-6736(14)60185-9

Lounibos LP, Kramer LD (2016) Invasiveness of *Aedes aegypti* and *Aedes albopictus* and vectorial capacity for Chikungunya virus. J Infect Dis 214(suppl 5):S453–S458. https://doi.org/10.1093/infdis/jiw285

Lulla A, Lulla V, Merits A (2012) Macromolecular assembly-driven processing of the 2/3 cleavage site in the alphavirus replicase polyprotein. J Virol 86(1):553–565. https://doi.org/10.1128/JVI.05195-11

Lulla V, Karo-Astover L, Rausalu K, Merits A, Lulla A (2013) Presentation overrides specificity: probing the plasticity of alphaviral proteolytic activity through mutational analysis. J Virol 87(18):10207–10220. https://doi.org/10.1128/JVI.01485-13

Lum FM, Teo TH, Lee WW, Kam YW, Renia L, Ng LF (2013) An essential role of antibodies in the control of Chikungunya virus infection. J Immunol 190(12):6295–6302. https://doi.org/10.4049/jimmunol.1300304

Lumsden WH (1955) An epidemic of virus disease in Southern province, Tanganyika territory, in 1952–53. II. General description and epidemiology. Trans R Soc Trop Med Hyg 49(1):33–57

Mahauad-Fernandez WD, Jones PH, Okeoma CM (2014) Critical role for bone marrow stromal antigen 2 in acute Chikungunya virus infection. J Gen Virol 95(Pt 11):2450–2461. https://doi.org/10.1099/vir.0.068643-0

Malet H, Coutard B, Jamal S, Dutartre H, Papageorgiou N, Neuvonen M, Ahola T, Forrester N, Gould EA, Lafitte D, Ferron F, Lescar J, Gorbalenya AE, de Lamballerie X, Canard B (2009) The crystal structures of Chikungunya and Venezuelan equine encephalitis virus nsP3 macro domains define a conserved adenosine binding pocket. J Virol 83(13):6534–6545. https://doi.org/10.1128/JVI.00189-09

Marimoutou C, Vivier E, Oliver M, Boutin JP, Simon F (2012) Morbidity and impaired quality of life 30 months after Chikungunya infection: comparative cohort of infected and uninfected French military policemen in Reunion Island. Medicine (Baltimore) 91(4):212–219. https://doi.org/10.1097/MD.0b013e318260b604

Mathur K, Anand A, Dubey SK, Sanan-Mishra N, Bhatnagar RK, Sunil S (2016) Analysis of Chikungunya virus proteins reveals that non-structural proteins nsP2 and nsP3 exhibit RNA interference (RNAi) suppressor activity. Sci Rep 6:38065. https://doi.org/10.1038/srep38065

McFarlane M, Arias-Goeta C, Martin E, O'Hara Z, Lulla A, Mousson L, Rainey SM, Misbah S, Schnettler E, Donald CL, Merits A, Kohl A, Failloux AB (2014) Characterization of *Aedes aegypti* innate-immune pathways that limit Chikungunya virus replication. PLoS Negl Trop Dis 8(7):e2994. https://doi.org/10.1371/journal.pntd.0002994

Messaoudi I, Vomaske J, Totonchy T, Kreklywich CN, Haberthur K, Springgay L, Brien JD, Diamond MS, Defilippis VR, Streblow DN (2013) Chikungunya virus infection results in higher and persistent viral replication in aged rhesus macaques due to defects in anti-viral immunity. PLoS Negl Trop Dis 7(7):e2343. https://doi.org/10.1371/journal.pntd.0002343

Miller JR, Koren S, Dilley KA, Puri V, Brown DM, Harkins DM, Thibaud-Nissen F, Rosen B, Chen X-G, Tu Z, Sharakhov IV, Sharakhova MV, Sebra R, Stockwell TB, Bergman NH, Sutton GG, Phillippy AM, Piermarini PM, Shabman RS (2017) Analysis of the <em> Aedes albopictus </em > C6/36 genome provides insight into cell line adaptations to < em > in vitro </em> viral propagation. bioRxiv. https://doi.org/10.1101/157081

Miner JJ, Aw-Yeang HX, Fox JM, Taffner S, Malkova ON, Oh ST, Kim AHJ, Diamond MS, Lenschow DJ, Yokoyama WM (2015) Chikungunya viral arthritis in the United States: a mimic of seronegative rheumatoid arthritis. Arthritis Rheumatol 67(5):1214–1220. https://doi.org/10.1002/art.39027

Nunes MR, Faria NR, de Vasconcelos JM, Golding N, Kraemer MU, de Oliveira LF, Azevedo Rdo S, da Silva DE, da Silva EV, da Silva SP, Carvalho VL, Coelho GE, Cruz AC, Rodrigues SG, Vianez JL Jr, Nunes BT, Cardoso JF, Tesh RB, Hay SI, Pybus OG, Vasconcelos PF (2015) Emergence and potential for spread of Chikungunya virus in Brazil. BMC Med 13:102. https://doi.org/10.1186/s12916-015-0348-x

Olagnier D, Scholte FE, Chiang C, Albulescu IC, Nichols C, He Z, Lin R, Snijder EJ, van Hemert MJ, Hiscott J (2014) Inhibition of dengue and Chikungunya virus infections by RIG-I-mediated type I interferon-independent stimulation of the innate antiviral response. J Virol 88(8):4180–4194. https://doi.org/10.1128/JVI.03114-13

Olson KE, Blair CD (2015) Arbovirus-mosquito interactions: RNAi pathway. Curr Opin Virol 15:119–126. https://doi.org/10.1016/j.coviro.2015.10.001

Olson KE, Higgs S, Hahn CS, Rice CM, Carlson JO, Beaty BJ (1994) The expression of chloramphenicol acetyltransferase in Aedes albopictus (C6/36) cells and Aedes triseriatus mosquitoes using a double subgenomic recombinant Sindbis virus. Insect Biochem Mol Biol 24(1):39–48

Ooi YS, Stiles KM, Liu CY, Taylor GM, Kielian M (2013) Genome-wide RNAi screen identifies novel host proteins required for alphavirus entry. PLoS Pathog 9(12):e1003835. https://doi.org/10.1371/journal.ppat.1003835

Panas MD, Ahola T, McInerney GM (2014) The C-terminal repeat domains of nsP3 from the Old World alphaviruses bind directly to G3BP. J Virol 88(10):5888–5893. https://doi.org/10.1128/JVI.00439-14

Panda D, Rose PP, Hanna SL, Gold B, Hopkins KC, Lyde RB, Marks MS, Cherry S (2013) Genome-wide RNAi screen identifies SEC61A and VCP as conserved regulators of Sindbis virus entry. Cell Rep 5(6):1737–1748. https://doi.org/10.1016/j.celrep.2013.11.028

Phillips A, Mossel E, Sanchez-Vargas I, Foy B, Olson K (2010) Alphavirus transducing system: tools for visualizing infection in mosquito vectors. J Vis Exp (45). https://doi.org/10.3791/2363

Pialoux G, Gauzere BA, Jaureguiberry S, Strobel M (2007) Chikungunya, an epidemic arbovirosis. Lancet Infect Dis 7(5):319–327. https://doi.org/10.1016/S1473-3099(07)70107-X

Pierro DJ, Myles KM, Foy BD, Beaty BJ, Olson KE (2003) Development of an orally infectious Sindbis virus transducing system that efficiently disseminates and expresses green fluorescent protein in *Aedes aegypti*. Insect Mol Biol 12(2):107–116

Poo YS, Nakaya H, Gardner J, Larcher T, Schroder WA, Le TT, Major LD, Suhrbier A (2014a) CCR2 deficiency promotes exacerbated chronic erosive neutrophil-dominated Chikungunya virus arthritis. J Virol 88(12):6862–6872. https://doi.org/10.1128/JVI.03364-13

Poo YS, Rudd PA, Gardner J, Wilson JA, Larcher T, Colle MA, Le TT, Nakaya HI, Warrilow D, Allcock R, Bielefeldt-Ohmann H, Schroder WA, Khromykh AA, Lopez JA, Suhrbier A (2014b) Multiple immune factors are involved in controlling acute and chronic Chikungunya virus infection. PLoS Negl Trop Dis 8(12):e3354. https://doi.org/10.1371/journal.pntd.0003354

Powers AM, Brault AC, Shirako Y, Strauss EG, Kang W, Strauss JH, Weaver SC (2001) Evolutionary relationships and systematics of the alphaviruses. J Virol 75(21):10118–10131. https://doi.org/10.1128/JVI.75.21.10118-10131.2001

Powers AM, Brault AC, Tesh RB, Weaver SC (2000) Re-emergence of Chikungunya and O'nyong-nyong viruses: evidence for distinct geographical lineages and distant evolutionary relationships. J Gen Virol 81(Pt 2):471–479. https://doi.org/10.1099/0022-1317-81-2-471

Priya R, Patro IK, Parida MM (2014) TLR3 mediated innate immune response in mice brain following infection with Chikungunya virus. Virus Res 189:194–205. https://doi.org/10.1016/j.virusres.2014.05.010

Pushko P, Geisbert J, Parker M, Jahrling P, Smith J (2001) Individual and bivalent vaccines based on alphavirus replicons protect guinea pigs against infection with Lassa and Ebola viruses. J Virol 75(23):11677–11685. https://doi.org/10.1128/JVI.75.23.11677-11685.2001

Rajapakse S, Rodrigo C, Rajapakse A (2010) Atypical manifestations of Chikungunya infection. Trans R Soc Trop Med Hyg 104(2):89–96. https://doi.org/10.1016/j.trstmh.2009.07.031

Rathore AP, Ng ML, Vasudevan SG (2013) Differential unfolded protein response during Chikungunya and Sindbis virus infection: CHIKV nsP4 suppresses eIF2alpha phosphorylation. Virol J 10:36. https://doi.org/10.1186/1743-422X-10-36

Rayner JO, Dryga SA, Kamrud KI (2002) Alphavirus vectors and vaccination. Rev Med Virol 12 (5):279–296. https://doi.org/10.1002/rmv.360

Renault P, Solet JL, Sissoko D, Balleydier E, Larrieu S, Filleul L, Lassalle C, Thiria J, Rachou E, de Valk H, Ilef D, Ledrans M, Quatresous I, Quenel P, Pierre V (2007) A major epidemic of Chikungunya virus infection on Reunion Island, France, 2005–2006. Am J Trop Med Hyg 77 (4):727–731

Rezza G, Nicoletti L, Angelini R, Romi R, Finarelli AC, Panning M, Cordioli P, Fortuna C, Boros S, Magurano F, Silvi G, Angelini P, Dottori M, Ciufolini MG, Majori GC, Cassone A, group Cs (2007) Infection with Chikungunya virus in Italy: an outbreak in a temperate region. Lancet 370 (9602):1840–1846. https://doi.org/10.1016/s0140-6736(07)61779-6

Rice CM, Levis R, Strauss JH, Huang HV (1987) Production of infectious RNA transcripts from Sindbis virus cDNA clones: mapping of lethal mutations, rescue of a temperature-sensitive marker, and in vitro mutagenesis to generate defined mutants. J Virol 61(12):3809–3819

Robinson MC (1955) An epidemic of virus disease in Southern Province, Tanganyika Territory, in 1952–53. I. Clinical features. Trans R Soc Trop Med Hyg 49 (1):28–32

Rudd PA, Wilson J, Gardner J, Larcher T, Babarit C, Le TT, Anraku I, Kumagai Y, Loo YM, Gale M Jr, Akira S, Khromykh AA, Suhrbier A (2012) Interferon response factors 3 and 7 protect against Chikungunya virus hemorrhagic fever and shock. J Virol 86(18):9888–9898. https://doi.org/10.1128/JVI.00956-12

Schilte C, Buckwalter MR, Laird ME, Diamond MS, Schwartz O, Albert ML (2012) Cutting edge: independent roles for IRF-3 and IRF-7 in hematopoietic and nonhematopoietic cells during host response to Chikungunya infection. J Immunol 188(7):2967–2971. https://doi.org/10.4049/jimmunol.1103185

Schilte C, Couderc T, Chretien F, Sourisseau M, Gangneux N, Guivel-Benhassine F, Kraxner A, Tschopp J, Higgs S, Michault A, Arenzana-Seisdedos F, Colonna M, Peduto L, Schwartz O, Lecuit M, Albert ML (2010) Type I IFN controls Chikungunya virus via its action on nonhematopoietic cells. J Exp Med 207(2):429–442. https://doi.org/10.1084/jem.20090851

Schilte C, Staikowsky F, Couderc T, Madec Y, Carpentier F, Kassab S, Albert ML, Lecuit M, Michault A (2013) Chikungunya virus-associated long-term arthralgia: a 36-month prospective longitudinal study. PLoS Negl Trop Dis 7(3):e2137. https://doi.org/10.1371/journal.pntd.0002137

Schuffenecker I, Iteman I, Michault A, Murri S, Frangeul L, Vaney MC, Lavenir R, Pardigon N, Reynes JM, Pettinelli F, Biscornet L, Diancourt L, Michel S, Duquerroy S, Guigon G, Frenkiel MP, Brehin AC, Cubito N, Despres P, Kunst F, Rey FA, Zeller H, Brisse S (2006) Genome microevolution of Chikungunya viruses causing the Indian Ocean outbreak. PLoS Med 3(7):e263. https://doi.org/10.1371/journal.pmed.0030263

Singh KR, Pavri KM (1967) Experimental studies with Chikungunya virus in *Aedes aegypti* and *Aedes albopictus*. Acta Virol 11(6):517–526

Souza-Neto JA, Sim S, Dimopoulos G (2009) An evolutionary conserved function of the JAK-STAT pathway in anti-dengue defense. Proc Natl Acad Sci U S A 106(42):17841–17846. https://doi.org/10.1073/pnas.0905006106

Steel JJ, Henderson BR, Lama SB, Olson KE, Geiss BJ (2011) Infectious alphavirus production from a simple plasmid transfection+. Virol J 8:356. https://doi.org/10.1186/1743-422X-8-356

Sun C, Gardner CL, Watson AM, Ryman KD, Klimstra WB (2014) Stable, high-level expression of reporter proteins from improved alphavirus expression vectors to track replication and dissemination during encephalitic and arthritogenic disease. J Virol 88(4):2035–2046. https://doi.org/10.1128/JVI.02990-13

Sy AK, Saito-Obata M, Medado IA, Tohma K, Dapat C, Segubre-Mercado E, Tandoc A 3rd, Lupisan S, Oshitani H (2016) Molecular characterization of Chikungunya virus, Philippines, 2011–2013. Emerg Infect Dis 22(5):887–890. https://doi.org/10.3201/eid2205.151268

Teng TS, Foo SS, Simamarta D, Lum FM, Teo TH, Lulla A, Yeo NK, Koh EG, Chow A, Leo YS, Merits A, Chin KC, Ng LF (2012) Viperin restricts Chikungunya virus replication and pathology. J Clin Invest 122(12):4447–4460. https://doi.org/10.1172/JCI63120

Teng TS, Kam YW, Lee B, Hapuarachchi HC, Wimal A, Ng LC, Ng LF (2015) A systematic meta-analysis of immune signatures in patients with acute Chikungunya virus infection. J Infect Dis 211(12):1925–1935. https://doi.org/10.1093/infdis/jiv049

Teo TH, Chan YH, Lee WW, Lum FM, Amrun SN, Her Z, Rajarethinam R, Merits A, Rotzschke O, Renia L, Ng LF (2017) Fingolimod treatment abrogates Chikungunya virus-induced arthralgia. Sci Transl Med 9(375). https://doi.org/10.1126/scitranslmed.aal1333

Teo TH, Lum FM, Claser C, Lulla V, Lulla A, Merits A, Renia L, Ng LF (2013) A pathogenic role for CD4 + T cells during Chikungunya virus infection in mice. J Immunol 190(1):259–269. https://doi.org/10.4049/jimmunol.1202177

Travanty EA, Adelman ZN, Franz AW, Keene KM, Beaty BJ, Blair CD, James AA, Olson KE (2004) Using RNA interference to develop dengue virus resistance in genetically modified *Aedes aegypti*. Insect Biochem Mol Biol 34(7):607–613. https://doi.org/10.1016/j.ibmb.2004.03.013

Tretyakova I, Hearn J, Wang E, Weaver S, Pushko P (2014) DNA vaccine initiates replication of live attenuated Chikungunya virus in vitro and elicits protective immune response in mice. J Infect Dis 209(12):1882–1890. https://doi.org/10.1093/infdis/jiu114

Tsetsarkin KA, McGee CE, Volk SM, Vanlandingham DL, Weaver SC, Higgs S (2009) Epistatic roles of E2 glycoprotein mutations in adaption of Chikungunya virus to *Aedes albopictus* and *Aedes aegypti* mosquitoes. PLoS One 4 (8):e6835. https://doi.org/10.1371/journal.pone.0006835

Tsetsarkin KA, Vanlandingham DL, McGee CE, Higgs S (2007) A single mutation in Chikungunya virus affects vector specificity and epidemic potential. PLoS Pathog 3(12): e201. https://doi.org/10.1371/journal.ppat.0030201

Tsetsarkin KA, Weaver SC (2011) Sequential adaptive mutations enhance efficient vector switching by Chikungunya virus and its epidemic emergence. PLoS Pathog 7(12):e1002412. https://doi.org/10.1371/journal.ppat.1002412

Utt A, Das PK, Varjak M, Lulla V, Lulla A, Merits A (2015) Mutations conferring a noncytotoxic phenotype on Chikungunya virus replicons compromise enzymatic properties of nonstructural protein 2. J Virol 89(6):3145–3162. https://doi.org/10.1128/JVI.03213-14

Utt A, Quirin T, Saul S, Hellstrom K, Ahola T, Merits A (2016) Versatile trans-replication systems for Chikungunya virus allow functional analysis and tagging of every replicase protein. PLoS ONE 11(3):e0151616. https://doi.org/10.1371/journal.pone.0151616

Van Bortel W, Dorleans F, Rosine J, Blateau A, Rousset D, Matheus S, Leparc-Goffart I, Flusin O, Prat C, Cesaire R, Najioullah F, Ardillon V, Balleydier E, Carvalho L, Lemaitre A, Noel H, Servas V, Six C, Zurbaran M, Leon L, Guinard A, van den Kerkhof J, Henry M, Fanoy E, Braks M, Reimerink J, Swaan C, Georges R, Brooks L, Freedman J, Sudre B, Zeller H (2014) Chikungunya outbreak in the Caribbean region, December 2013 to March 2014, and the significance for Europe. Euro Surveill 19(13)

Varghese FS, Thaa B, Amrun SN, Simarmata D, Rausalu K, Nyman TA, Merits A, McInerney GM, Ng LF, Ahola T (2016) The antiviral alkaloid berberine reduces Chikungunya virus-induced mitogen-activated protein kinase signaling. J Virol 90(21):9743–9757. https://doi.org/10.1128/JVI.01382-16

Vazeille M, Moutailler S, Coudrier D, Rousseaux C, Khun H, Huerre M, Thiria J, Dehecq JS, Fontenille D, Schuffenecker I, Despres P, Failloux AB (2007) Two Chikungunya isolates from the outbreak of La Reunion (Indian Ocean) exhibit different patterns of infection in the mosquito. Aedes albopictus. PLoS One 2(11):e1168. https://doi.org/10.1371/journal.pone.0001168

Waldock J, Olson KE, Christophides GK (2012) Anopheles gambiae antiviral immune response to systemic O'nyong-nyong infection. PLoS Negl Trop Dis 6(3):e1565. https://doi.org/10.1371/journal.pntd.0001565

Walker T, Jeffries CL, Mansfield KL, Johnson N (2014) Mosquito cell lines: history, isolation, availability and application to assess the threat of arboviral transmission in the United Kingdom. Parasit Vectors 7:382. https://doi.org/10.1186/1756-3305-7-382

Weaver SC, Forrester NL (2015) Chikungunya: Evolutionary history and recent epidemic spread. Antiviral Res 120:32–39. https://doi.org/10.1016/j.antiviral.2015.04.016

Weaver SC, Osorio JE, Livengood JA, Chen R, Stinchcomb DT (2012) Chikungunya virus and prospects for a vaccine. Expert Rev Vaccines 11(9):1087–1101. https://doi.org/10.1586/erv.12.84

Werneke SW, Schilte C, Rohatgi A, Monte KJ, Michault A, Arenzana-Seisdedos F, Vanlandingham DL, Higgs S, Fontanet A, Albert ML, Lenschow DJ (2011) ISG15 is critical in the control of Chikungunya virus infection independent of UbE1L mediated conjugation. PLoS Pathog 7(10):e1002322. https://doi.org/10.1371/journal.ppat.1002322

Wielanek AC, Monredon JD, Amrani ME, Roger JC, Serveaux JP (2007) Guillain-Barre syndrome complicating a Chikungunya virus infection. Neurology 69(22):2105–2107. https://doi.org/10.1212/01.wnl.0000277267.07220.88

Xi Z, Ramirez JL, Dimopoulos G (2008) The *Aedes aegypti* toll pathway controls dengue virus infection. PLoS Pathog 4(7):e1000098. https://doi.org/10.1371/journal.ppat.1000098

Chronic Chikungunya Virus Disease

Mary K. McCarthy, Bennett J. J. Davenport
and Thomas E. Morrison

Contents

Abstract Chikungunya virus (CHIKV) is a mosquito-transmitted alphavirus that has caused both small- and large-scale epidemics of incapacitating musculoskeletal disease across the globe. A substantial proportion of infected individuals experience debilitating arthralgia and/or arthritis that can persist in relapsing or continuous forms for months to years, an occurrence that appears independent of viral strain and outbreak location. Due to the lack of CHIKV-specific vaccine or therapeutics, treatment of chronic CHIKV disease is limited to supportive care. Although the

---

Mary K. McCarthy, Bennett J. J. Davenport—These authors contributed equally.

M. K. McCarthy · B. J. J. Davenport · T. E. Morrison (✉)
Department of Immunology and Microbiology, University of Colorado School of Medicine, Aurora, CO 80045, USA
e-mail: Thomas.Morrison@ucdenver.edu

Current Topics in Microbiology and Immunology (2022) 435: 55–80
https://doi.org/10.1007/82_2018_147

Published online: 18 January 2019

epidemiologic and molecular mechanisms that dictate resolution or chronicity of CHIKV disease remain unclear, several risk factors and immunological responses have been implicated in the development of chronic CHIKV disease. Mounting evidence from animal models and limited case studies indicates that chronic disease is likely a result of induced autoimmunity and/or viral persistence in joint-associated tissue. Due to the global spread and explosive, often unpredictable nature of CHIKV epidemics, concerted efforts to obtain a more precise understanding of the development and maintenance of chronic CHIKV disease must be at the forefront of investigative endeavors.

1 Introduction

Chikungunya virus (CHIKV) is a mosquito-borne alphavirus in the family *Togaviridae*. Although sylvatic cycles involving non-human primates and forest-dwelling mosquitoes have been identified in regions of East and West Africa, CHIKV is capable of sustained epidemic transmission between *Aedes* species mosquitoes (*Aedes aegypti* and *Aedes albopictus*) and humans. CHIKV re-emerged in the mid-2000s to cause large outbreaks of acute rheumatologic disease worldwide. While many individuals fully recover, a high proportion of infected individuals experience chronic incapacitating arthralgia, which can last for months to years.

2 What Is Chronic CHIKV Disease?

Acute CHIKV infection generally presents with a sudden onset of a high fever, severe joint pain, muscle aches, and rash, which subside in 7–10 days. Severe, excruciating pain in the joints (arthralgia) is a hallmark symptom of CHIKV infection and is a dominating aspect of the post-acute stage. Post-acute and chronic stages of CHIKV musculoskeletal disease are typically defined as a failure to return to the preexisting condition 3 weeks to 3 months or more than 3 months, respectively, beyond the onset of acute disease.

3 When and Where Has Chronic Disease Associated with CHIKV Infection Been Observed?

Historical accounts of numerous febrile disease outbreaks in the late 1700s, 1800s, and early 1900s that could almost certainly be attributed to CHIKV infection universally describe severe protracted joint pain, particularly in the small joints of the hands and feet, lasting for several weeks to months after the cessation of fever

(Carey 1971). These outbreaks occurred in Indonesia, Tanzania, Egypt, Burma, China, and India and highlight that protracted joint pain has been a consistent feature of CHIKV disease since the earliest descriptions.

CHIKV was first isolated from patient serum samples in 1953 during an outbreak of acute febrile illness on the Makonde Plateau in what is now Tanzania (Ross 1956). During this outbreak, it was noted that recurring joint pains occurred "intermittently in the majority of patients and in some continued up to 4 months after the original illness" (Robinson 1955). The second published report regarding a small CHIKV outbreak in South Africa in 1956 reported on the prolonged convalescence and recurring joint pains as characteristic features of the cases (Gear and Reid 1957).

Large-scale outbreaks occurred in Thailand and India in the 1960s and 1970s (Shah et al. 1964; Myers et al. 1965; Halstead et al. 1969a, b). Reports from these epidemics focus mainly on epidemiology and acute disease signs, with little documented follow-up of patients after acute disease. However, anecdotal evidence from the 1964 CHIKV epidemic in southern India suggested chronic symptoms were an important component of CHIKV disease, where it was reported that a number of patients experienced "persistent, intermittently incapacitating joint pains for several months following their illness, and some still had mild residual pains a year later" (Carey et al. 1969). Another study from the same outbreak reported persistent or recurrent arthralgia "for a not inconsiderable number of patients" for weeks to months (De Ranitz et al. 1965). CHIKV seemingly disappeared from India in 1973. However, other Southeast Asian countries have reported frequent but irregular outbreaks from the 1960s onward (Powers and Logue 2007).

Since its isolation, CHIKV outbreaks have occurred throughout Africa. It is from some of these outbreaks in the mid-1970s that the frequency of chronic symptoms following acute CHIKV infection was first quantified. Chronic musculoskeletal disease was reported in a group of 28 individuals who contracted CHIKV in South Africa—five of these patients continued to suffer from arthritis and episodes of fever 20 months later (Fourie and Morrison 1979). In an early clinical description of acute and chronic CHIKV infection, Kennedy et al. noted the persistence of articular pain, tenderness, swelling, and functional impairment for at least 4–6 months, with symptoms gradually improving in two individuals who were followed for 16 months (Kennedy et al. 1980). In a larger retrospective study of 107 patients who were infected in South Africa in 1975–1977, 5.6% of patients reported persistent joint pain and stiffness 3–5 years after acute infection (Brighton et al. 1983). In the ensuing years, CHIKV was responsible for outbreaks and sporadic cases in several African countries, with a particularly large outbreak of nearly 50,000 estimated cases in the Congo in 1999–2000 (Moore et al. 1974; Tomori et al. 1975; Saluzzo et al. 1983; Muyembe-Tamfum et al. 2003; Powers and Logue 2007).

In 2004, a CHIKV outbreak began on Lamu Island off the coast of Kenya (Sergon et al. 2008). The virus subsequently spread to the Comoros, predominantly the island of Grande Comore, affecting nearly 200,000 people (Sergon et al. 2007), and La Reunion Island, where it is estimated to have infected nearly 40% of that island's population of 785,000 (Renault et al. 2007). Neighboring Indian Ocean islands, including Madagascar, Mayotte, Seychelles, and Mauritius, were also hard hit during this epidemic (Powers and Logue 2007; Sissoko et al. 2008). At the end of 2005, CHIKV was reported in India for the first time in over 30 years where it resulted in millions of disease cases (Ravi 2006). The virus continued to spread to Sri Lanka and many other countries in Southeast Asia, with suspected cases ultimately numbering in the millions (Petersen and Powers 2016). During this time, viremic travelers returning from epidemic regions established autochthonous transmission cycles in CHIKV-naïve areas, including France and Italy (Moro et al. 2010; Grandadam et al. 2011). The first locally acquired cases of CHIKV in the Western Hemisphere were reported on the Caribbean island of St. Martin in December 2013 (Leparc-Goffart et al. 2014). Since then, the virus has spread to over 45 countries in the Caribbean, Central America, and South America. While the number of cases in most of the Americas has recently waned, the epidemic continues in Brazil, with over 120,000 confirmed autochthonous transmission cases in 2017 alone (www.paho.org/chikungunya). Remarkably, the development of chronic musculoskeletal disease signs and symptoms remains a consistent feature across all of these recent epidemics, seemingly independent of viral strain, geography, and population.

4 Frequency of Chronic CHIKV Disease Manifestations

With the expansive CHIKV outbreaks in the Indian Ocean, Southeast Asia, and the Americas in the last 10–15 years, the occurrence and frequency of chronic disease signs and symptoms have received increased attention. The long-term sequela of CHIKV infection has been summarized in a recent review by van Aalst and colleagues (van Aalst et al. 2017). Persistent polyarthralgia and/or arthritis are the most commonly reported chronic disease manifestations. Other signs and symptoms that have been reported less frequently include fatigue, headache, alopecia, depression, sleep disorders, memory impairment, hearing difficulties, blurred vision, digestive complaints, and skin lesions/rashes (Gérardin et al. 2011; Yaseen et al. 2014; Marimoutou et al. 2015; Bouquillard et al. 2017).

The most comprehensive reports of chronic CHIKV disease have come from patients infected on Reunion Island or from French travellers returning from the Indian Ocean region during the 2005–2007 epidemic. Depending on the time of follow-up, the frequency of self-reported persistent arthralgia among these patients is generally between 40 and 60% (Table 1). Subsequent outbreaks in India, Southeast Asia, and Italy have also described chronic CHIKV disease occurring in frequencies (range 1.6–67%) comparable to that of Reunion Island (Table 1). Initial reports from the most recent outbreak in the Americas suggest a similar rate of persistent arthralgia (range 42–89%) among infected patients with 6–12 months follow-up times (Table 1) (Rodríguez-Morales et al. 2016b, a; Murillo-Zamora et al. 2017; Rodriguez-Morales et al. 2017). While the proportion of patients

Table 1 Frequency of chronic musculoskeletal disease manifestations quantified from outbreaks in the indicated regions

Region	Outbreak location	Outbreak Year(s)	Number of patients	Diagnosis/inclusion criteria	Follow-up time	Percent with chronic symptoms	Reference
Africa	South Africa	1975	28 patients	Symptoms, serology, viral culture	18 months	17.9% complained of episodic polyarthritis	Fourie and Morrison (1979)
	South Africa	1975–1977	107 patients	Serology	3–5 years	12% not fully recovered	Brighton et al. (1983)
Indian Ocean	Reunion Island	2005–2006	221 individuals	Questionnaire; CHIKV+ if they thought they got infected	12–15 months	55.6% reported relapses; 10% had chronic	Staikowsky et al. (2008)
	Reunion Island	2005–2006	88 patients (66% hospitalized for acute CHIKV)	Serology or RT-PCR	18 months	63.6% reported persistent arthralgia; continuous in 55.4% of the persistent patients	Borgherini et al. (2008)
	Reunion Island	2005–2006	147 patients	Serology or RT-PCR	15 months	57% reported rheumatic symptoms (of those: 63% permanent trouble and 37% relapsing)	Sissoko et al. (2009)
	Reunion Island	2006	1094 subjects	Serology	18 months	43% of seropositive reported musculoskeletal pain and other symptoms	Gérardin et al. (2011)
	Reunion Island	2006	382 subjects	Questionnaire; CHIKV+ if they thought they got infected	30 months	36.7% of CHIKV+ considered themselves not healed	Marimoutou et al. (2012)
	Reunion Island	2005–2006	146 subjects	Serology, declared musculoskeletal pain at disease onset	2 years (average)	31.1% reported relapsing musculoskeletal pain; 43.3% reported lingering musculoskeletal pain	Gérardin et al. (2013)
	Reunion Island	2005–2006	180 patients	In hospital for febrile arthralgia; CHIKV IgG tested but not excluded if negative	4–36 months	~60% reported long-term arthralgia	Schilte et al. (2013)

(continued)

Table 1 (continued)

Region	Outbreak location	Outbreak Year(s)	Number of patients	Diagnosis/inclusion criteria	Follow-up time	Percent with chronic symptoms	Reference
	Reunion Island	2005–2006	307 patients	Serology or epidemiological case definition	32 months (mean)	83.1% reported persistent joint pain but functional impairment was moderate	Bouquillard et al. (2017)
	Indian Ocean	2005–2006	47 travelers from Indian Ocean region	Serology, RT-PCR, and/or viral culture	1, 3, and 6 months	88% after 1 month, 86% after 3 months, and 48% after 6 months	Simon et al. (2007)
	Indian Ocean	2005–2006	29 travelers from Indian Ocean region	Serology or RT-PCR	2 years	17 (59%) still suffered from persistent arthralgia	Larrieu et al. (2010)
	Indian Ocean	2005–2007	391 travelers from Indian Ocean region	Serology or RT-PCR	23.4 months (median)	55% considered not recovered at 1 year	Couturier et al. (2012)
	Mauritius	2006	173 subjects	Questionnaire	27.5 months	78.6% reported persisting musculoskeletal symptoms	Essackjee et al. (2013)
India and Southeast Asia	India	2008–2009	203 patients	Serologically confirmed	1–9 months	51% cure rate at the end of 9 months	Manimunda et al. (2010)
	India	2009	1195 individuals	Serology or epidemiological case definition	18 months	48.6% of affected individuals had persistent pain	Rahim et al. (2016)
	India	2007	437 cases	Serology or epidemiological case definition; no prior rheumatologic disease	15 months	57% had post-viral non-specific polyarthralgia; 22% inflammatory polyarthritis; 27.7% soft tissue rheumatism	Mathew et al. (2011)
	India	2006	509 cases	Serology	1–2 years	4.1% had persistent rheumatic pains at 1 year; 1.6% at 2 years	Chopra et al. (2012)
	Singapore	2008	97 patients	RT-PCR during acute phase	6 weeks	35.9% had persistent arthralgia	Win et al. (2010)
	Sri Lanka	2006	1001 subjects	Community survey	12–36 months	56% had persisting disability which lasted 12, 24, and 36 months in 17.8, 9.5, and 6.1%, respectively	Kularatne et al. (2012)
	Malaysia	2008–2009	53 patients	Serology, RT-PCR, and/or viral culture	15–24 months	45% had persistent arthralgia beyond 4 months; 22.5% had arthralgia at one year	Zim et al. (2013)

(continued)

Table 1 (continued)

Region	Outbreak location	Outbreak Year(s)	Number of patients	Diagnosis/inclusion criteria	Follow-up time	Percent with chronic symptoms	Reference
Europe	Italy	2007	250 patients	Serology or included if possible case	12 months	66.5% reported myalgia, asthenia, or arthralgia	Moro et al. (2012)
Americas	Colombia	2014–2015	39 patients	RT-PCR during acute phase	35 weeks (median)	89.7% reported persistent polyarthralgia	Rodríguez-Morales et al. (2016a)
	Colombia	2015	131 patients	Serology	12 months	44.3% with persistent polyarthralgia	Rodríguez-Morales et al. (2016b)
	Mexico	2014–2015	136 subjects	Serology	6 months	41.9% reported persistent arthralgia	Murillo-Zamora et al. (2017)
	Colombia	2015	111 cases	Validated WHO criteria; no other arboviral infections; no prior rheumatologic disease	12 months	45.6% reported persistent rheumatological symptoms	Rodriguez-Morales et al. (2017)
	Curacao	2015	304 patients	Serology or RT-PCR	400 days	64% defined as chronically affected	Elsinga et al. (2017)

experiencing chronic disease wanes over time, impaired quality of life has been reported even up to six years after CHIKV infection (Marimoutou et al. 2015).

High rates of musculoskeletal pain are reported in healthy CHIKV seronegative populations (8–17% (Gérardin et al. 2011; Mathew et al. 2011; Marimoutou et al. 2015; Feldstein et al. 2017), highlighting the importance of being able to differentiate between persistent arthralgia that is directly attributable to CHIKV infection or that is unrelated to the CHIKV infection. This is often done by excluding patients who reported arthralgia or who were diagnosed with a rheumatologic disorder prior to CHIKV infection. A few studies have made direct comparisons between infected individuals and uninfected case-control populations. In one study, persistent musculoskeletal pains in CHIKV seropositive individuals from Reunion Island approximately 2 years after infection were found to be highly attributable to CHIKV (Gérardin et al. 2011). In fact, CHIKV infection explained nearly two-thirds of the musculoskeletal pains reported by seropositive individuals and approximately 30% of musculoskeletal pain reported in the general population. In a recent report from the US Virgin Islands, CHIKV seropositive case patients had an almost three-fold increased risk for persistent arthralgia compared with non-symptomatic controls (Feldstein et al. 2017). Thus, persistent rheumatic symptoms reported in infected individuals can be largely attributed to CHIKV infection.

5 Disease Signs and Symptoms of Post-acute CHIKV Disease

While most reports have focused on the prevalence of post-CHIKV chronic symptoms, relying primarily on patient questionnaires and physical examination, recent reports have focused on the clinical expression (Manimunda et al. 2010; Mathew et al. 2011) and systematic categorization of chronic CHIKV diseases. Long-term clinical follow-up of patients infected during the 2005–2006 CHIKV outbreak on Reunion Island who had no evidence of a rheumatologic disorder prior to CHIKV infection has led to the categorization of two main types of chronic CHIKV disease: post-CHIKV musculoskeletal disorders (MSDs) and post-CHIKV de novo chronic inflammatory rheumatisms (CIRs) (Javelle et al. 2015; Simon et al. 2015). Post-CHIKV de novo CIRs include rheumatoid arthritis (RA), spondyloarthritis (SA), and undifferentiated polyarthritis (UP). Patients who do not fulfill the American College of Rheumatology/European League Against Rheumatism (ACR/EULAR) criteria for RA (Aletaha et al. 2010) or the European Spondyloarthropathy Study Group (ESSG) criteria for SA (Dougados et al. 1991) are evaluated for other common causes of polyarthritis, such as gout, chronic viral hepatitis, and sarcoidosis. If these causes are eliminated and no alternative diagnosis is present, the patient is diagnosed with UP. Patients with persistent musculoskeletal pain, but no arthritis are diagnosed with post-CHIKV MSDs, which can be further categorized as loco-regional (generally affecting previously injured joints) or diffuse (generally distal polyarthralgia involving the hands and/or feet).

A few reports have differentiated between relapsing/episodic and chronic/continuous arthralgia. Although the follow-up time and definition of relapse varies, these studies have suggested high rates of both relapsing (21–50% of all patients) and continuous (10–55%) arthralgia (Staikowsky et al. 2008; Borgherini et al. 2008; Gérardin et al. 2013; Schilte et al. 2013; Essackjee et al. 2013). In one report, 72% of patients with confirmed CHIKV infection reported relapses, defined as a recurrence of joint pain after one-week of being symptom-free, within the first two years after infection (Couturier et al. 2012). The intensity of joint pain experienced during a relapse can be equal to or more severe than pain during the acute phase of infection and generally affects the same joints.

In some cases, radiographic or ultrasonographic imaging has been employed to further characterize the manifestations of chronic CHIKV disease. The most common findings within 6 months of disease onset are symmetrical multiple tenosynovitis affecting the wrists and ankles (Simon et al. 2007). Mogami and colleagues performed extensive ultrasonographic analysis of patients with no history of rheumatologic disease who had chronic musculoskeletal symptoms ~4 months from disease onset during the recent CHIKV outbreak in Brazil (Mogami et al. 2017a, c). The most common ultrasonographic findings in these patients were synovitis and tenosynovitis of the small joints in the wrists, fingers, and ankles. They also reported joint effusion, cellulitis, and myositis. In patients examined an average of 10 months after CHIKV disease onset, Manimunda and colleagues reported joint effusion, bone erosion, marrow edema, synovial thickening, tendonitis, and tenosynovitis (Manimunda et al. 2010). A portion of patients who had joint pain prior to CHIKV infection showed joint degeneration by MRI and X-ray, suggesting that CHIKV could have exacerbated a preexisting rheumatologic disorder. Finally, in one study of 94 patients who were free of any disease prior to CHIKV infection and who met the criteria for post-CHIKV de novo CIRs, nearly half of these patients had bone lesions detectable by radiography at a median follow-up time of 3.5 years post CHIKV infection (Javelle et al. 2015).

Collectively, this work has revealed that there is a wide spectrum of de novo post-CHIKV musculoskeletal diseases and that proper classification of post-CHIKV conditions and exclusion of differential diagnoses is important in order to provide the most efficient and effective treatment.

6 Risk Factors for Chronic CHIKV Disease

Risk factors for the development of chronic CHIKV disease have been identified by multiple studies, although several studies are either in conflict or lack the numerical power to be able to identify statistically significant associations (Table 2). Anecdotal observations from the 1970s noted that younger individuals (<17 years of age) made up a high proportion of the cases who fully recovered (Brighton et al.

Table 2 Risk factors associated with the development of chronic CHIKV disease

Reference	Risk factor(s) identified	Factors not significantly associated
Couturier et al. (2012)	Age > 50 years, underlying arthritis or presence of at least one comorbidity, longer acute stage with joint swelling	Female gender, fever
Essackjee et al. (2013)	Age > 50 years, female gender, joint stiffness and swelling in acute phase, symmetrical distribution of joint symptoms during acute phase	
Gérardin et al. (2013)	Age > 45 years, severe rheumatic involvement at acute phase, CHIKV-specific IgG titer at time of analysis	Female gender
Hoarau et al. (2010)	Age > 60 years, higher viral loads during acute phase, high CRP at D0	
Larrieu et al. (2010)	Severity and duration of initial phase (30 days or more with severe pain)	Female gender, age, comorbidity
Moro et al. (2010)	Age, female gender, and history of rheumatologic disease	IgM positivity at 4–5 months or 12–13 months, IgG median titer at 12–13 months
Murillo-Zamora et al. (2017)	Age > 40 years, articular pain at 3 months post disease onset	Female gender, severity and duration of initial pain, associated signs and symptoms, underlying chronic disease
Rahim et al. (2016)	Female gender, history of RMSD, history of joint swelling during acute stage, joint involvement in persistent pain, and vegetarian diet	
Ramachandran et al. (2014)	Age > 35 years, high-grade fever, involvement of 4 or more types of joints, joint swelling; females had longer duration of joint pain	
Schilte et al. (2013)	Age > 35 years, high C-reactive protein at D0, diabetes at D0, presence of arthralgia, memory disorder, and/or concentration disorder 4 months after disease onset	Female gender, viral load, hospitalization during acute phase
Sissoko et al. (2009)	Age > 45 years, severe initial joint pain, presence of underlying osteoarthritis	
Win et al. (2010)	Female gender, lower peak creatinine level	Peak viral load, viremia, and duration of fever during acute phase
Yaseen et al. (2014)	Age, > 4 days of sick leave, initial severity of joint swelling, depressed mood at acute stage and 6 months after disease onset	
Zim et al. (2013)	Age > 40 years	Female gender, arthritis at acute phase, number of joints involved, pain score at acute phase, preexisting rheumatologic condition, thrombocytopenia, leukopenia

1983). These observations have been corroborated by more recent studies, which indicate that increasing age (variably identified as >35, >45, >50, or >60 years of age) is a significant independent risk factor for the development of persistent rheumatologic symptoms following CHIKV disease (Sissoko et al. 2009; Hoarau et al. 2010; Couturier et al. 2012; Moro et al. 2012; Gérardin et al. 2013; Zim et al. 2013; Schilte et al. 2013; Essackjee et al. 2013; Yaseen et al. 2014). The role of age as a risk factor for chronic disease has also been implicated in animal models of CHIKV infection. Viral RNA can be readily detected in the spleen of aged non-human primates (NHP) 35 days following CHIKV infection yet remains undetectable in young adult controls, and aged mice show elevated viral RNA levels in joint-associated tissues when compared with adult controls at 60 days post infection (Messaoudi et al. 2013; Uhrlaub et al. 2016).

In some studies, the severity of initial joint pain or swelling and longer duration of the acute stage was associated with the development of chronic symptoms (Sissoko et al. 2009; Larrieu et al. 2010; Couturier et al. 2012; Yaseen et al. 2014). Similarly, there is some evidence that the presence of rheumatologic symptoms at 4 or 6 months after acute disease onset can be predictive of chronic rheumatologic disease remaining one to two years later (Schilte et al. 2013; Yaseen et al. 2014). Female gender is an independent risk factor for post-CHIKV chronic disease (Win et al. 2010; Couturier et al. 2012; Essackjee et al. 2013; Rahim et al. 2016). Comorbidities, especially a history of rheumatologic disease, are additional risk factors for chronic CHIKV disease (Sissoko et al. 2009; Couturier et al. 2012; Moro et al. 2012; Rahim et al. 2016). CHIKV has been hypothesized to act as a trigger for autoimmune disorders such as RA and SA (discussed more below), but may also exacerbate preexisting rheumatologic conditions.

Gérardin and colleagues reported a significant positive association between CHIKV-specific IgG titers and chronic CHIKV disease (defined as ongoing musculoskeletal pain), with the lowest levels of CHIKV-specific IgG reported in the group considered to be recovered, higher levels in the group with relapsing disease, and the highest levels in the group with lingering disease (Gérardin et al. 2013). However, other studies comparing CHIKV-specific IgM or IgG levels in chronic and recovered patients failed to confirm these findings (Hoarau et al. 2010, 2013). Thus, any possible associations between ongoing chronic CHIKV disease and levels of CHIKV-specific IgG require further study. There is some evidence that the timing of anti-CHIKV antibody responses during the acute phase may play a role in recovery. In one study, high levels of CHIKV-specific IgG3 appearing early during the acute phase of infection (and potentially triggered by high viral loads) were associated with recovery, whereas the late appearance of CHIKV-specific IgG3 was associated with the development of chronic symptoms (Kam et al. 2012).

Additional associations with chronic CHIKV disease have been identified by single studies and remain to be confirmed by additional reports or in larger patient cohorts. In a longitudinal study of 30 patients with laboratory-confirmed CHIKV infection, individuals who developed persistent arthralgia had higher serum levels of IL-6 and granulocyte macrophage-colony stimulating factor (GM-CSF) during the post-acute stage than those who recovered, while recovery was associated with high levels of eotaxin and hepatocyte growth factor (HGF) (Chow et al. 2011). However, a separate study did not detect different levels of IL-6 in acute phase serum of patients who ultimately recovered or those who developed chronic arthralgia (Hoarau et al. 2010).

In terms of virologic factors, a report of 49 hospitalized patients from Reunion Island suggested that higher viral loads during the acute phase of infection may be a risk factor for the development of chronic disease (Hoarau et al. 2010), although this association was not confirmed by other groups (Win et al. 2010; Schilte et al. 2013). Sequencing of viral isolates from patient serum during acute disease is rarely performed. In one study based on partial sequencing of the CHIKV E1 gene in 10 viremic patients, viral amino acid complexity at the acute stage was associated with an increased number of painful joints and intensity of sequelae at day 300 (Thiberville et al. 2013). Although the sample size was small and these findings bear further study, this study represents the first attempt to define a potential relationship between CHIKV genetic diversity during acute infection and disease outcome.

In summary, many potential risk factors for chronic CHIKV disease have been identified and studies involving larger patient cohorts are necessary to confirm previously observed associations or identify new risk factors. While causal relationships between these risk factors and chronic disease have yet to be defined, animal models of CHIKV infection provide testable venues for many of the observations. Further study may identify opportunities for early intervention to prevent the development of chronic disease in high-risk patients.

7 Immunological Characteristics of Chronic CHIKV Disease

An understanding of the immunobiology of chronic CHIKV disease has been impeded by the difficulty in accessing tissues relevant to persistent disease, such as synovial tissue and other musculoskeletal tissues. Thus, most studies have interrogated serum and peripheral blood mononuclear cells (PBMCs) with only a select few characterizing joint-associated tissues. Although the characterization of immune processes associated with chronic CHIKV disease remains sparse, some aspects of the immune response in naïve or recovered versus chronic disease patients can be differentiated.

7.1 T Cells

In a select few studies, immunological responses in muscle and synovial tissues of patients with chronic CHIKV disease have been described. Analysis of a quadriceps muscle biopsy from a patient presenting with chronic CHIKV symptoms revealed substantial leukocyte infiltrates (Ozden et al. 2007). Among the more prominent cell types detected were $CD3^+$ T cells, although further stratification of this cell subset, such as their CD4 or CD8 positivity, was not performed (Ozden et al. 2007). Similar to these observations, immunological characterization of perivascular synovial tissue collected from a patient with chronic CHIKV disease revealed immune cell infiltrates that included activated CD4 and CD8 T cells (as shown by HLA-DR and CD69 expression), suggesting that activated T cells are present within joint-associated tissue during chronic CHIKV disease (Hoarau et al. 2010).

A few studies have analyzed peripheral T cells in patients with chronic CHIKV disease. Using CyTOF analysis, one group found that the frequency of activated CD8 T cells in the blood was elevated in patients with chronic CHIKV disease compared with healthy controls (Miner et al. 2015). In a different study, CHIKV-specific T cell responses in patients that had recovered from acute CHIKV disease were compared to those in patients that developed chronic CHIKV disease. Examination of peripheral blood T cell responses to CHIKV-specific antigens was assessed by measuring IFNγ production in response to CHIKV peptide pools derived from overlapping 20 amino acid peptide sequences spanning CHIKV E2, capsid, and nsP1 (Hoarau et al. 2013). Although the intensity of responses was not different between recovered and chronic CHIKV disease patients, these analyses revealed that all patients who recovered from acute CHIKV disease (6/6 patients, 100%) responded to E2 peptides, whereas only a portion of chronic CHIKV disease patients (16/27 patients, $\sim$59%) exhibited T cell reactivity to E2 peptides (Hoarau et al. 2013). In addition, in recovered patients E2 was the most frequently recognized antigen, whereas in chronic patients nsP1 was the most frequently recognized antigen. This shift in immunoreactivity of antiviral T cell responses between patients that had recovered and those that developed chronic disease suggests an alteration in T cell immunodominance. This study also suggests that global T cell responses against CHIKV E2 may influence the outcome of CHIKV disease.

Finally, one study suggests CD4 T regulatory (Treg) cells may be dysregulated during chronic CHIKV disease. In patients with acute CHIKV infection, peripheral CD4 Treg cells are decreased in frequency compared with naïve controls and returned to normal levels in those patients that did not progress to chronic disease. In contrast, Treg numbers remained low in patients that developed chronic CHIKV symptoms (Kulkarni et al. 2017). However, a separate report found that the frequency and activation status of peripheral CD4 Treg cells were equivalent in patients with acute CHIKV infection and healthy controls (Wauquier et al. 2011). Thus, characterizing the role of CD4 Treg cells during both acute and chronic CHIKV disease requires further investigation.

7.2 NK Cells

The role of natural killer (NK) cells in CHIKV disease has largely been investigated during the acute or convalescent phases of disease, and only a select few reports have characterized NK cells during chronic disease. Evaluation of synovial fluid cells from a patient with chronic CHIKV disease demonstrated the presence of activated $CD69^{+}$ NK cells, whereas circulating NK cells from the same patient appeared quiescent with nearly undetectable CD69 expression (Hoarau et al. 2010). Examination of a muscle tissue biopsy revealed the presence of NK cells in a patient suffering from chronic CHIKV disease, although their numbers were reportedly very low in comparison to other immune cell subsets within the tissue (Ozden et al. 2007). Although limited by the number of patients and tissues analyzed, these studies suggest that musculoskeletal tissues can be infiltrated with NK cells during chronic CHIKV disease.

The phenotype and function of NK cells in the circulation of patients with chronic CHIKV disease, in comparison with healthy controls, also have been evaluated. In one study, the frequency of NK cells in PBMCs was elevated in chronic CHIKV disease patients compared with healthy controls (Miner et al. 2015). In more detailed analyses, a separate study found that, following in vitro stimulation, the frequency of NK cells expressing perforin and CD107a was reduced in PBMCs from chronic CHIKV disease patients compared with healthy controls, suggesting that reduced NK cell cytolytic activity is associated with chronic disease (Thanapati et al. 2017). In contrast, the frequency of NK cells expressing TNFα and IFNγ was increased compared with healthy controls, suggesting that elevated proinflammatory cytokine production by NK cells may contribute to chronic CHIKV disease (Shegarfi et al. 2012; Thanapati et al. 2017). Taken together, these findings show that NK cells are engaged in the context of chronic CHIKV disease, although whether NK cell expansion and activation is a bystander of chronic disease, or whether dysregulated NK cell subsets exacerbate chronic CHIKV disease remains to be determined.

7.3 *Monocytes and Macrophages*

Immunohistochemistry performed on both muscle and synovial tissue biopsies from patients with chronic CHIKV disease detected $CD68^{+}$ and $CD18^{+}$ cells, respectively, suggesting that musculoskeletal tissues of patients with chronic CHIKV disease are infiltrated with macrophages (Ozden et al. 2007; Hoarau et al. 2010). Consistent with these data, abundant $CD14^{+}$ cells and ballooned macrophages were detected in the synovial fluid collected from the same patient as the synovial tissue biopsy (Hoarau et al. 2010). Remarkably, a subset of the macrophages in the synovial fluid and the tissue biopsy stained positive for CHIKV antigen (Hoarau et al. 2010). Importantly, neither cellular infiltrates nor CHIKV antigen was detected in similar specimens obtained from patients that had recovered from acute CHIKV disease (Hoarau et al. 2010). Although limited to two patients, these data suggest that chronic CHIKV disease is associated with accumulation of monocytes and macrophages in musculoskeletal tissues, and a subset of these cells appears to harbor viral antigen (Ozden et al. 2007; Hoarau et al. 2010).

7.4 B Cells and Antibody

Patients that have either recovered from CHIKV-induced symptoms or develop chronic CHIKV disease display antibody class switch and generation of a neutralizing CHIKV-specific antibody response (Kam et al. 2012; Hoarau et al. 2013). However, the phenotypic and functional properties of CHIKV-specific B cells associated with recovery or the development of chronic CHIKV disease remain poorly defined. During the acute stage of infection, peripheral blood B cells exhibit an activated phenotype, which correlates with the generation of CHIKV-specific antibody responses (Hoarau et al. 2010).

A few studies have performed comparative analyses of anti-CHIKV antibody responses between patients that recovered from acute CHIKV disease and those that developed persistent symptoms. As discussed above, the delayed generation of anti-CHIKV IgG3 neutralizing antibody during acute infection was associated with the development of chronic CHIKV disease, suggesting that the timing or quality of the neutralizing antibody response can influence the development of chronic disease (Kam et al. 2012). Consistent with these data, studies in mice have shown that pathogenic CHIKV strains establish persistent infection in joint-associated tissues by evading humoral immunity and more severe chronic disease in aged animals is associated with defects in the development of the anti-CHIKV neutralizing antibody response (Hawman et al. 2016; Uhrlaub et al. 2016).

7.5 Cytokines and Chemokines

Cytokine and chemokine profiling of CHIKV-infected patients have identified some associations with chronic CHIKV disease (Table 3). In one study, both recovered and chronic CHIKV disease patients had elevated serum IL-12 levels during the acute phase. However, at later times IL-12 levels in recovered patients returned to baseline, whereas IL-12 remained elevated in patients with chronic CHIKV disease up to 12 months following acute infection (Hoarau et al. 2010). Further analysis during the chronic phase revealed that IFNα expression was elevated in PBMCs of patients with chronic CHIKV disease compared with PBMCs from patients that had recovered from the disease (Hoarau et al. 2010). Patients with chronic CHIKV disease also have been found to have increased circulating levels of IL-6 and GM-CSF and decreased levels of eotaxin and HGF at 3 months post infection, in comparison with patients that recovered from the disease (Chow et al. 2011). Other cytokine and immunological dysregulation include elevated IL-1α, IL-15, and complement component C3 at 36 months post infection in chronic disease patients compared with recovered controls (Schilte et al. 2013). CCL2, IL-6, and IL-8 have been detected in synovial fluid collected from a patient with chronic CHIKV disease, yet these factors were undetectable in the serum of the same patient (Hoarau et al. 2010). These findings suggest that CCL2 expression within joint-associated tissue, and decreased circulating levels of the CCR2 antagonist eotaxin, may foster

Table 3 Cytokine, chemokine, and immune factor characteristics in chronic CHIKV disease

Cytokine/chemokine	Characteristics in chronic disease	Tissue characterized	Comparator	Time of analysis post infection	Reference (s)
IL-1α	Elevated	Serum	Recovered patients	36 months	Schilte et al. (2013)
IL-6	Elevated	Serum	Recovered patients	2–3 months; 6 months	Chow et al. (2011), Kelvin et al. (2011)
	Present	Synovial fluid	n/a	18 months	Hoarau et al. (2010)
IL-8	Present	Synovial fluid	n/a	18 months	Hoarau et al. (2010)
IL-12	Elevated	Serum	Recovered patients	3–12 months	Hoarau et al. (2010)
IL-15	elevated	serum	recovered patients	36 months	Schilte et al. (2013)
IFNα	Elevated	PBMC	Recovered patients	12 months	Hoarau et al. (2010)
GM-CSF	Elevated	Serum	recovered patients	2–3 months	Chow et al. (2011)
Eotaxin	Decreased	Serum	Recovered patients	2–3 months	Chow et al. (2011)
Hepatocyte growth factor	Decreased	Serum	Recovered patients	2–3 months	Chow et al. (2011)
CCL2	Present	Synovial fluid	n/a	18 months	Hoarau et al. (2010)
CXCL9	Elevated	Serum	Asymptomatic patients	6 months	Kelvin et al. (2011)
CXCL10	elevated	serum	asymptomatic patients	6 months	Kelvin et al. (2011)
C3	Elevated	Serum	Recovered patients	36 months	Schilte et al. (2013)

recruitment of monocyte cell populations to joint-associated tissues during chronic CHIKV disease (Ogilvie et al. 2001; Hoarau et al. 2010; Chow et al. 2011). Additional dysregulated chemotactic pathways identified in chronic CHIKV disease include elevations of the T cell and monocyte chemoattractants CXCL9 and CXCL10 in serum at 6 months post infection, (Kelvin et al. 2011; Chopra et al. 2014). Collectively, these studies suggest that specific cytokines and chemokines are elevated during chronic CHIKV disease; however, additional analysis in patient cohorts and mechanistic studies in animal models are required to improve an understanding of their contributions to the development, duration, and/or severity of chronic CHIKV disease.

8 Therapeutic Interventions for Chronic CHIKV Disease

Due to the absence of a vaccine or specific therapeutics, clinical treatment for patients with post-acute and chronic CHIKV disease is limited to supportive care. However, two recent reports described comprehensive recommendations for the treatment of both acute and chronic CHIKV disease (Simon et al. 2015; Marques et al. 2017).

For patients with post-acute CHIKV disease, treatments initiated during the acute phase, such as administration of analgesics, can be maintained so long as they remain efficacious against the patient symptoms. Common analgesics used in this setting include paracetamol or dipyrone. For cases in which analgesics fail to provide clinical benefit, and diagnosis has ruled out dengue fever, non-steroidal anti-inflammatory drugs (NSAIDs) such as naproxen, celecoxib, or etoricoxib are often prescribed. Treatment with NSAIDs is typically maintained for several weeks. In cases with moderate to severe musculoskeletal pain and inflammatory presentations (tenosynovitis, synovitis) that do not respond to NSAIDs, corticosteroids such as prednisone or prednisolone are the recommended alternatives. In severe cases with prolonged chronic symptoms despite treatment with analgesics and NSAIDs, such as patients that develop post-CHIKV CIRs, disease-modifying anti-rheumatic drugs (DMARDs) may be utilized, and commonly include methotrexate (MTX) and/or sulfasalazine (SSZ). Biological DMARDs, such as anti-TNF blockers, although safe and effective in the treatment of RA, are only recommended in patients that present with persistent inflammatory joint disease after other treatment options have proven ineffective. As mentioned earlier, post-CHIKV CIRs can be further classified as RA, SA, or UP. For patients exhibiting post-CHIKV RA, administration of DMARDs constitutes the first line of treatment, with biological DMARDs being considered only as a final alternative. In cases of post-CHIKV SA, NSAIDs represent the first treatment option, and DMARDs are prescribed only as a second line of treatment when patients are refractory to NSAIDs. Lastly, for patients diagnosed with post-CHIKV UP, DMARDs are employed only in situations in which NSAIDs and corticosteroids are ineffective.

Although results in clinical trials have not provided unequivocal affirmation of efficacy, antimalarials such as chloroquine or hydroxychloroquine have been utilized as monotherapy or in combination with other treatments. Hydroxychloroquine as a monotherapy, although effective in alleviating chronic CHIKV symptoms, was not any more efficacious than DMARD monotherapy (Chopra et al. 2014). Evidence in one exploratory clinical trial suggested that hydroxychloroquine in combination with DMARDs yields positive prognosis in patients with chronic disease (Chopra et al. 2014; Ravindran and Alias 2017).

In summary, treatment of persistent symptoms following CHIKV infection requires careful evaluation and monitoring by clinicians. New, rigorous clinical trials in patient populations with distinct disease states are needed to provide guidance for physicians and to improve clinical management of post-acute and chronic stages of CHIKV disease.

9 Mechanisms of Chronic Disease Pathogenesis

The underlying mechanisms of chronic CHIKV disease are not well understood, although the induction of autoimmunity and/or viral persistence in joint tissue have been hypothesized as major contributory factors. However, chronic CHIKV disease could also be due to unresolved injury in joint tissue either caused or exacerbated by acute infection. Evidence for autoimmunity and viral persistence is discussed in more detail below.

9.1 Autoimmunity

Acute CHIKV infection has been hypothesized to act as a trigger for the development of RA in susceptible individuals. This is supported by the overlapping gene signature observed in mouse models of acute CHIKV infection and collagen-induced arthritis (Nakaya et al. 2012), and similar blood T cell subset profiles in CHIKV and RA patients compared with healthy controls (Miner et al. 2015). The majority of studies have reported zero or low prevalence of autoimmune markers such as rheumatoid factor (RF), antinuclear antibodies (ANA), and anti-citrullinated protein antibodies (ACPA) in patients with chronic CHIKV disease (Manimunda et al. 2010; Schilte et al. 2013; Waymouth et al. 2013; Chopra et al. 2014; Venugopalan et al. 2014; Foissac et al. 2015; Miner et al. 2015; Blettery et al. 2016; Mogami et al. 2017b). However, a few studies have reported on a subset of patients who developed symptoms that met the ACR criteria for RA following acute CHIKV infection. When measured, the prevalence of ACPA and/or RF in these patients ranged from 5 to 66% (Chopra et al. 2008; Bouquillard and Combe 2009; Manimunda et al. 2010; Ganu and Ganu 2011; Javelle et al. 2015). The low prevalence of RF and ACPA in some studies may reflect a shorter follow-up time, or it may simply indicate that post-CHIKV RA is largely seronegative. In studies that directly compared recovered and chronic disease patients, no association between ANA levels and disease was detected (Schilte et al. 2013; Venugopalan et al. 2014).

While one report did not observe classic RA-like erosions in hands or feet of patients diagnosed with post-CHIKV chronic inflammatory arthritis (Chopra et al. 2008), these patients were only being followed within 6 months of acute disease onset and X-ray was used to detect joint erosions, which is not as sensitive as MRI. Other studies with an extended follow-up time of 2–5 years after acute disease report a high prevalence ($\sim$80%) of joint damage in patients diagnosed with post-CHIKV de novo RA (Bouquillard and Combe 2009; Javelle et al. 2015), suggesting that it takes time to accumulate sufficient joint damage to be visible in radiographic or MRI analysis.

Comprehensive analysis of larger cohorts directly comparing post-CHIKV RA and CHIKV-naïve RA populations is required to derive more concrete conclusions about the pathogenesis of the two conditions. Regardless of the differential etiology, the management of classic RA and post-CHIKV RA are quite similar, with DMARDs (especially MTX) being the first-line treatment of choice (Javelle et al. 2015). While it is clear that a subset of patients develops RA or an RA-like disease after CHIKV infection, the induction of autoimmunity (in genetically predisposed individuals or arising de novo) likely cannot explain the high prevalence of post-CHIKV rheumatologic disorders.

9.2 Persistence of Viral Replication, Viral PAMPs, and/or Viral Antigen

Chronic virus infection can lead to persistent inflammation and injury of infected tissue. Thus, persistent CHIKV infection has long been hypothesized to be a contributing factor to chronic CHIKV disease. However, due to the invasive nature and difficulty in obtaining samples from joints, particularly small joints of the wrists, hands, and feet, tissues of patients with chronic CHIKV disease are not routinely tested for the presence of infectious virus, viral RNA, or viral antigen. In rare cases, synovial fluid or synovial biopsies collected from patients diagnosed with chronic CHIKV disease have been collected to test for the presence of viral RNA, with the results being negative in most cases, although details regarding the quantity of fluid collected and how the assays were performed are lacking (Ganu and Ganu 2011; Bouquillard et al. 2017). However, when immunohistochemistry is used to detect viral antigen in biopsy samples, infected cells have been readily identified. Muscle biopsies from two different patients (either 10 days or 3 months after acute disease onset) showed the presence of CHIKV antigen in muscle satellite cells (Ozden et al. 2007). In addition, CHIKV antigen and CHIKV RNA were detected in perivascular synovial macrophages and synovial biopsy tissue of a patient with chronic CHIKV disease 18 months after acute disease onset, but not in the synovial biopsies of two recovered patients (Hoarau et al. 2010).

More definitive evidence of CHIKV persistence comes from experiments using NHPs and mice. In WT immunocompetent mice, CHIKV RNA persists uniquely in joint-associated tissues for at least 16 weeks post infection (Hawman et al. 2013, 2016). In addition, both positive- and negative-strand CHIKV RNA can be detected in the feet of adult WT C57BL/6 mice infected with CHIKV out to at least 100 days post infection (Poo et al. 2014), and in mice infected with a recombinant luciferase-expressing CHIKV strain, luciferase activity can be detected at 60 days post infection (Teo et al. 2013). Finally, CHIKV RNA and low levels of infectious virus were detected in feet of adult and aged mice at 60 and 90 days post infection (Uhrlaub et al. 2016). The presence of negative-strand viral RNA and ongoing reporter gene activity suggests that the persistent state is marked by low levels of productive viral replication within affected tissues. In cynomolgus macaques, joints, muscles, lymphoid organs, and liver have been shown to harbor infectious CHIKV or CHIKV RNA for weeks post inoculation (Labadie et al. 2010). CHIKV RNA also persists in the spleen of aged rhesus macaques at 5 weeks post infection (Messaoudi et al. 2013). Together, these findings indicate that persistence of CHIKV RNA and/or antigen in lymphoid and joint-associated tissues is a common feature of CHIKV infection, even in immunocompetent hosts.

In mice, persistence of viral RNA in joint-associated tissues was associated with chronic synovitis (Hawman et al. 2013), suggesting that viral persistence drives chronic inflammation. Additionally, $Rag1^{-/-}$ mice persistently infected with CHIKV had higher viral burdens and developed more severe signs of chronic inflammation in joint-associated tissues compared with WT mice, suggesting that functional adaptive immune responses protect against, rather than exacerbate, chronic musculoskeletal disease (Hawman et al. 2013). Consistent with these data, the presence of CHIKV antigen in the synovium of a patient with chronic arthralgia was associated with high levels of apoptosis, synovial tissue injury, and type I IFN expression (Hoarau et al. 2010). In addition, elevated levels of IFNα were detected in PBMCs isolated from patients with chronic CHIKV disease (Hoarau et al. 2010). This evidence suggests that persistent viral replication or persistence of viral products fosters sustained immune activation and injury in joint-associated tissue. While very low levels of infectious virus and the ongoing presence of negative-strand RNA suggests that CHIKV likely continues to replicate during persistent infection, continuous production of new viral particles may not be necessary to drive chronic disease. Persistence of viral RNA alone may provide sufficient activation of TLRs or cytosolic RNA sensors to promote prolonged immune activation and chronic disease in joint tissues (McCarthy and Morrison 2017).

Acknowledgements This work was supported by Public Health Service grants R01 AI108725, R01 AI123348, and U19 AI109680 (to T. E. Morrison), F32 AI122463 (to M. K. McCarthy), and T32 AR007534 (to B. J. J. Davenport) from the National Institute of Allergy and Infectious Diseases.

References

Aletaha D, Neogi T, Silman AJ, Funovits J, Felson DT, Bingham CO, Birnbaum NS, Burmester GR, Bykerk VP, Cohen MD, Combe B, Costenbader KH, Dougados M, Emery P, Ferraccioli G, Hazes JMW, Hobbs K, Huizinga TWJ, Kavanaugh A, Kay J, Kvien TK, Laing T, Mease P, Ménard HA, Moreland LW, Naden RL, Pincus T, Smolen JS, Stanislawska-Biernat E, Symmons D, Tak PP, Upchurch KS, Vencovský J, Wolfe F, Hawker G (2010) 2010 Rheumatoid arthritis classification criteria: an American College of Rheumatology/European League against rheumatism collaborative initiative. Arthritis Rheum 62:2569–2581. https://doi.org/10.1002/art.27584

Blettery M, Brunier L, Polomat K, Moinet F, Deligny C, Arfi S, Jean-Baptiste G, De Bandt M (2016) Brief report: management of chronic post-chikungunya rheumatic disease: the Martinican experience: chronic post-chikungunya rheumatic disease. Arthritis Rheumatol 68:2817–2824. https://doi.org/10.1002/art.39775

Borgherini G, Poubeau P, Jossaume A, Gouix A, Cotte L, Michault A, Arvin-Berod C, Paganin F (2008) Persistent arthralgia associated with chikungunya virus: a study of 88 adult patients on Reunion Island. Clin Infect Dis 47:469–475. https://doi.org/10.1086/590003

Bouquillard É, Combe B (2009) A report of 21 cases of rheumatoid arthritis following chikungunya fever. A mean follow-up of two years. Joint Bone Spine 76:654–657. https://doi.org/10.1016/j.jbspin.2009.08.005

Bouquillard E, Fianu A, Bangil M, Charlette N, Ribéra A, Michault A, Favier F, Simon F, Flipo R-M (2017) Rheumatic manifestations associated with Chikungunya virus infection: A study of 307 patients with 32-month follow-up (Rhumatochik study). Joint Bone Spine. https://doi.org/10.1016/j.jbspin.2017.01.014

Brighton SW, Prozesky OW, de la Harpe AL (1983) Chikungunya virus infection. A retrospective study of 107 cases. South Afr Med J Suid-Afr Tydskr Vir Geneeskd 63:313–315

Carey DE (1971) Chikungunya and dengue: a case of mistaken identity? J Hist Med Allied Sci 26:243–262

Carey DE, Myers RM, DeRanitz CM, Jadhav M, Reuben R (1969) The 1964 chikungunya epidemic at Vellore, South India, including observations on concurrent dengue. Trans R Soc Trop Med Hyg 63:434–445

Chopra A, Anuradha V, Ghorpade R, Saluja M (2012) Acute Chikungunya and persistent musculoskeletal pain following the 2006 Indian epidemic: a 2-year prospective rural community study. Epidemiol Infect 140:842–850. https://doi.org/10.1017/S0950268811001300

Chopra A, Anuradha V, Lagoo-Joshi V, Kunjir V, Salvi S, Saluja M (2008) Chikungunya virus aches and pains: an emerging challenge. Arthritis Rheum 58:2921–2922. https://doi.org/10.1002/art.23753

Chopra A, Saluja M, Venugopalan A (2014) Effectiveness of chloroquine and inflammatory cytokine response in patients with early persistent musculoskeletal pain and arthritis following chikungunya virus infection. Arthritis Rheumatol 66:319–326. https://doi.org/10.1002/art.38221

Chow A, Her Z, Ong EKS, Chen J, Dimatatac F, Kwek DJC, Barkham T, Yang H, Rénia L, Leo Y-S, Ng LFP (2011) Persistent arthralgia induced by chikungunya virus infection is associated with interleukin-6 and granulocyte macrophage colony-stimulating factor. J Infect Dis 203:149–157. https://doi.org/10.1093/infdis/jiq042

Couturier E, Guillemin F, Mura M, Léon L, Virion J-M, Letort M-J, De Valk H, Simon F, Vaillant V (2012) Impaired quality of life after chikungunya virus infection: a 2-year follow-up study. Rheumatology 51:1315–1322. https://doi.org/10.1093/rheumatology/kes015

De Ranitz CM, Myers RM, Varkey MJ, Isaac ZH, Carey DE (1965) Clinical impressions of chikungunya in Vellore gained from study of adult patients. Indian J Med Res 53:756–763

Dougados M, Linden SVD, Juhlin R, Huitfeldt B, Amor B, Calin A, Cats A, Dijkmans B, Olivieri I, Pasero G, Veys E, Zeidler H (1991) The European spondyloarthropathy study group preliminary criteria for the classification of spondylarthropathy. Arthritis Rheum 34:1218–1227. https://doi.org/10.1002/art.1780341003

Elsinga J, Gerstenbluth I, van der Ploeg S, Halabi Y, Lourents NT, Burgerhof JG, van der Veen HT, Bailey A, Grobusch MP, Tami A (2017) Long-term Chikungunya Sequelae in Curaçao: Burden, Determinants, and a Novel Classification Tool. J Infect Dis 216:573–581. https://doi.org/10.1093/infdis/jix312

Essackjee K, Goorah S, Ramchurn SK, Cheeneebash J, Walker-Bone K (2013) Prevalence of and risk factors for chronic arthralgia and rheumatoid-like polyarthritis more than 2 years after infection with chikungunya virus. Postgrad Med J 89:440–447. https://doi.org/10.1136/postgradmedj-2012-131477

Feldstein LR, Rowhani-Rahbar A, Staples JE, Weaver MR, Halloran ME, Ellis EM (2017) Persistent arthralgia associated with chikungunya virus outbreak, US Virgin Islands, December 2014–February 2016. Emerg Infect Dis 23:673–676. https://doi.org/10.3201/eid2304.161562

Foissac M, Javelle E, Ray S, Guerin B, Simon F (2015) Post-chikungunya rheumatoid arthritis, saint martin. Emerg Infect Dis 21:530–532. https://doi.org/10.3201/eid2103.141397

Fourie ED, Morrison JG (1979) Rheumatoid arthritic syndrome after chikungunya fever. South Afr Med J 56:130–132

Ganu MA, Ganu AS (2011) Post-chikungunya chronic arthritis—our experience with DMARDs over two year follow up. J Assoc Physicians India 59:83–86

Gear J, Reid FP (1957) The occurrence of a dengue-like fever in the North-Eastern Transvaal. I. Clinical features and isolation of virus. South Afr Med J Suid-Afr Tydskr Vir Geneeskd 31:253–257

Gérardin P, Fianu A, Malvy D, Mussard C, Boussaïd K, Rollot O, Michault A, Gaüzere B-A, Bréart G, Favier F (2011) Perceived morbidity and community burden after a chikungunya outbreak: the TELECHIK survey, a population-based cohort study. BMC Med 9:5. https://doi.org/10.1186/1741-7015-9-5

Gérardin P, Fianu A, Michault A, Mussard C, Boussaïd K, Rollot O, Grivard P, Kassab S, Bouquillard E, Borgherini G, Gaüzère B-A, Malvy D, Bréart G, Favier F (2013) Predictors of chikungunya rheumatism: a prognostic survey ancillary to the TELECHIK cohort study. Arthritis Res Ther 15:R9. https://doi.org/10.1186/ar4137

Grandadam M, Caro V, Plumet S, Thiberge J-M, Souarès Y, Failloux A-B, Tolou HJ, Budelot M, Cosserat D, Leparc-Goffart I, Desprès P (2011) Chikungunya virus, Southeastern France. Emerg Infect Dis 17:910–913. https://doi.org/10.3201/eid1705.101873

Halstead SB, Scanlon JE, Umpaivit P, Udomsakdi S (1969a) Dengue and chikungunya virus infection in man in Thailand, 1962–1964. IV. Epidemiologic studies in the Bangkok Metropolitan area. Am J Trop Med Hyg 18:997–1021. https://doi.org/10.4269/ajtmh.1969.18.997

Halstead SB, Udomsakdi S, Scanlon JE, Rohitayodhin S (1969b) Dengue and chikungunya virus infection in man in Thailand, 1962–1964. V. Epidemiologic observations outside Bangkok. Am J Trop Med Hyg 18:1022–1033. https://doi.org/10.4269/ajtmh.1969.18.1022

Hawman DW, Stoermer KA, Montgomery SA, Pal P, Oko L, Diamond MS, Morrison TE (2013) Chronic joint disease caused by persistent chikungunya virus infection is controlled by the adaptive immune response. J Virol 87:13878–13888. https://doi.org/10.1128/JVI.02666-13

Hawman DW, Fox JM, Ashbrook AW, May NA, Schroeder KMS, Torres RM, Crowe JE Jr, Dermody TS, Diamond MS, Morrison TE (2016) Pathogenic chikungunya virus evades B cell Responses to establish persistence. Cell Rep 16:1326–1338. https://doi.org/10.1016/j.celrep.2016.06.076

Hoarau JJ, Jaffar Bandjee MC, Krejbich Trotot P, Das T, Li-Pat-Yuen G, Dassa B, Denizot M, Guichard E, Ribera A, Henni T, Tallet F, Moiton MP, Gauzere BA, Bruniquet S, Jaffar Bandjee Z, Morbidelli P, Martigny G, Jolivet M, Gay F, Grandadam M, Tolou H, Vieillard V, Debre P, Autran B, Gasque P (2010) Persistent chronic inflammation and infection by chikungunya arthritogenic alphavirus in spite of a robust host immune response. J Immunol 184:5914–5927. https://doi.org/10.4049/jimmunol.0900255

Hoarau J-J, Gay F, Pellé O, Samri A, Jaffar-Bandjee M-C, Gasque P, Autran B (2013) Identical strength of the T cell responses against E2, nsP1 and capsid CHIKV proteins in recovered and chronic patients after the epidemics of 2005–2006 in La Reunion Island. PLoS ONE 8:e84695

Javelle E, Ribera A, Degasne I, Gaüzère B-A, Marimoutou C, Simon F (2015) Specific management of post-chikungunya rheumatic disorders: a retrospective study of 159 cases in Reunion Island from 2006–2012. PLoS Negl Trop Dis 9:e0003603. https://doi.org/10.1371/journal.pntd.0003603

Kam Y-W, Simarmata D, Chow A, Her Z, Teng T-S, Ong EKS, Rénia L, Leo Y-S, Ng LFP (2012) Early appearance of neutralizing immunoglobulin G3 antibodies is associated with chikungunya virus clearance and long-term clinical protection. J Infect Dis 205:1147–1154. https://doi.org/10.1093/infdis/jis033

Kelvin AA, Banner D, Silvi G, Moro ML, Spataro N, Gaibani P, Cavrini F, Pierro A, Rossini G, Cameron MJ, Bermejo-Martin JF, Paquette SG, Xu L, Danesh A, Farooqui A, Borghetto I, Kelvin DJ, Sambri V, Rubino S (2011) Inflammatory cytokine expression is associated with chikungunya virus resolution and symptom severity. PLoS Negl Trop Dis 5:e1279. https://doi.org/10.1371/journal.pntd.0001279

Kennedy AC, Fleming J, Solomon L (1980) Chikungunya viral arthropathy: a clinical description. J Rheumatol 7:231–236

Kularatne SAM, Weerasinghe SC, Gihan C, Wickramasinghe S, Dharmarathne S, Abeyrathna A, Jayalath T (2012) Epidemiology, Clinical Manifestations, and Long-Term Outcomes of a

Major Outbreak of Chikungunya in a Hamlet in Sri Lanka, in 2007: A Longitudinal Cohort Study. J Trop Med 2012:1–6. https://doi.org/10.1155/2012/639178
Kulkarni SP, Ganu M, Jayawant P, Thanapati S, Ganu A, Tripathy AS (2017) Regulatory T cells and IL-10 as modulators of chikungunya disease outcome: a preliminary study. Eur J Clin Microbiol Infect Dis, 1–7. https://doi.org/10.1007/s10096-017-3087-4
Labadie K, Larcher T, Joubert C, Mannioui A, Delache B, Brochard P, Guigand L, Dubreil L, Lebon P, Verrier B, de Lamballerie X, Suhrbier A, Cherel Y, Le Grand R, Roques P (2010) Chikungunya disease in nonhuman primates involves long-term viral persistence in macrophages. J Clin Invest 120:894–906. https://doi.org/10.1172/JCI40104
Larrieu S, Pouderoux N, Pistone T, Filleul L, Receveur M-C, Sissoko D, Ezzedine K, Malvy D (2010) Factors associated with persistence of arthralgia among chikungunya virus-infected travellers: report of 42 French cases. J Clin Virol 47:85–88. https://doi.org/10.1016/j.jcv.2009.11.014
Leparc-Goffart I, Nougairede A, Cassadou S, Prat C, de Lamballerie X (2014) Chikungunya in the Americas. The Lancet 383:514. https://doi.org/10.1016/S0140-6736(14)60185-9
Manimunda SP, Vijayachari P, Uppoor R, Sugunan AP, Singh SS, Rai SK, Sudeep AB, Muruganandam N, Chaitanya IK, Guruprasad DR (2010) Clinical progression of chikungunya fever during acute and chronic arthritic stages and the changes in joint morphology as revealed by imaging. Trans R Soc Trop Med Hyg 104:392–399. https://doi.org/10.1016/j.trstmh.2010.01.011
Marimoutou C, Ferraro J, Javelle E, Deparis X, Simon F (2015) Chikungunya infection: self-reported rheumatic morbidity and impaired quality of life persist 6 years later. Clin Microbiol Infect 21:688–693. https://doi.org/10.1016/j.cmi.2015.02.024
Marimoutou C, Vivier E, Oliver M, Boutin J-P, Simon F (2012) Morbidity and Impaired Quality of Life 30 Months After Chikungunya Infection: Comparative Cohort of Infected and Uninfected French Military Policemen in Reunion Island. Medicine 91:212–219. https://doi.org/10.1097/MD
Marques CDL, Duarte ALBP, Ranzolin A, Dantas AT, Cavalcanti NG, Gonçalves RSG, Junior LF da R, Valadares LD de A, Melo AKG de, Freire EAM, Teixeira R, Neto FAB, Medeiros MM das C, Carvalho JF de, Santos MSF, Océa RA de LC, Levy RA, Andrade CAF de, Pinheiro G da RC, Abreu MM, Verztman JF, Merenlender S, Ribeiro SLE, Costa IP da, Pileggi G, Trevisani VFM, Lopes MIB, Brito C, Figueiredo E, Queiroga F, Feitosa T, Tenório A da S, Siqueira GR de, Paiva R, Vasconcelos JTS, Christopoulos G (2017) Recommendations of the Brazilian Society of Rheumatology for the diagnosis and treatment of chikungunya fever. Part 2—Treatment. Rev Bras Reumatol Engl Ed. https://doi.org/10.1016/j.rbre.2017.06.004
Mathew AJ, Goyal V, George E, Thekkemuriyil DV, Jayakumar B, Chopra A, On behalf of the Trivandrum COPCORD Study Group (2011) Rheumatic-musculoskeletal pain and disorders in a naïve group of individuals 15 months following a chikungunya viral epidemic in south India: a population based observational study. Int J Clin Pract 65:1306–1312. https://doi.org/10.1111/j.1742-1241.2011.02792.x
McCarthy MK, Morrison TE (2017) Persistent RNA virus infections: do PAMPS drive chronic disease? Curr Opin Virol 23:8–15. https://doi.org/10.1016/j.coviro.2017.01.003
Messaoudi I, Vomaske J, Totonchy T, Kreklywich CN, Haberthur K, Springgay L, Brien JD, Diamond MS, DeFilippis VR, Streblow DN (2013) Chikungunya virus infection results in higher and persistent viral replication in aged Rhesus macaques due to defects in anti-viral immunity. PLoS Negl Trop Dis 7:e2343. https://doi.org/10.1371/journal.pntd.0002343
Miner JJ, Aw Yeang HX, Fox JM, Taffner S, Malkova ON, Oh ST, Kim AHJ, Diamond MS, Lenschow DJ, Yokoyama WM (2015) Brief report: chikungunya viral arthritis in the United States: a mimic of seronegative rheumatoid arthritis. Arthritis Rheumatol 67:1214–1220. https://doi.org/10.1002/art.39027
Mogami R, de Almeida Vieira A, Junqueira Filho EA, Lopes AJ (2017a) Chikungunya fever outbreak in Rio de Janeiro, Brazil: ultrasonographic aspects of musculoskeletal complications. J Clin Ultrasound 45:43–44. https://doi.org/10.1002/jcu.22394

Mogami R, Pereira Vaz JL, de Fátima Barcelos Chagas Y, de Abreu MM, Torezani RS, de Almeida Vieira A, Junqueira Filho EA, Barbosa YB, Carvalho ACP, Lopes AJ (2017b) Ultrasonography of hands and wrists in the diagnosis of complications of chikungunya fever. J Ultrasound Med. https://doi.org/10.1002/jum.14344

Mogami R, Vaz JLP, Chagas Y de FB, Torezani RS, Vieira A de A, Koifman ACB, Barbosa YB, de Abreu MM (2017c) Ultrasound of ankles in the diagnosis of complications of chikungunya fever. Radiol Bras 50:71–75. https://doi.org/10.1590/0100-3984.2016.0221

Moore DL, Reddy S, Akinkugbe FM, Lee VH, David-West TS, Causey OR, Carey DE (1974) An epidemic of chikungunya fever at Ibadan, Nigeria, 1969. Ann Trop Med Parasitol 68:59–68

Moro ML, Gagliotti C, Silvi G, Angelini R, Sambri V, Rezza G, Massimiliani E, Mattivi A, Grilli E, Finarelli AC, Spataro N, Pierro AM, Seyler T, Macini P (2010) Chikungunya virus in North-Eastern Italy: a seroprevalence survey. Am J Trop Med Hyg 82:508–511. https://doi.org/10.4269/ajtmh.2010.09-0322

Moro ML, Grilli E, Corvetta A, Silvi G, Angelini R, Mascella F, Miserocchi F, Sambo P, Finarelli AC, Sambri V, Gagliotti C, Massimiliani E, Mattivi A, Pierro AM, Macini P (2012) Long-term chikungunya infection clinical manifestations after an outbreak in Italy: A prognostic cohort study. J Infect 65:165–172. https://doi.org/10.1016/j.jinf.2012.04.005

Murillo-Zamora E, Mendoza-Cano O, Trujillo-Hernández B, Alberto Sánchez-Piña R, Guzmán-Esquivel J (2017) Persistent arthralgia and related risks factors in laboratory-confirmed cases of chikungunya virus infection in Mexico. Rev Panam Salud Pública 41

Muyembe-Tamfum JJ, Peyrefitte CN, Yogolelo R, Mathina Basisya E, Koyange D, Pukuta E, Mashako M, Tolou H, Durand JP (2003) Epidemic of chikungunya virus in 1999 and 200 in the democratic republic of the Congo. Med Trop Rev Corps Sante Colon 63:637–638

Myers RM, Carey DE, Reuben R, Jesudass ES, De Ranitz C, Jadhav M (1965) The 1964 epidemic of dengue-like fever in South India: isolation of chikungunya virus from human sera and from mosquitoes. Indian J Med Res 53:694–701

Nakaya HI, Gardner J, Poo Y-S, Major L, Pulendran B, Suhrbier A (2012) Gene profiling of chikungunya virus arthritis in a mouse model reveals significant overlap with rheumatoid arthritis. Arthritis Rheum 64:3553–3563. https://doi.org/10.1002/art.34631

Ogilvie P, Bardi G, Clark-Lewis I, Baggiolini M, Uguccioni M (2001) Eotaxin is a natural antagonist for CCR2 and an agonist for CCR5. Blood 97:1920–1924. https://doi.org/10.1182/blood.V97.7.1920

Ozden S, Huerre M, Riviere J-P, Coffey LL, Afonso PV, Mouly V, de Monredon J, Roger J-C, Amrani ME, Yvin J-L, Jaffar M-C, Frenkiel M-P, Sourisseau M, Schwartz O, Butler-Browne G, Desprès P, Gessain A, Ceccaldi P-E (2007) Human muscle satellite cells as targets of chikungunya virus infection. PLoS ONE 2:e527. https://doi.org/10.1371/journal.pone.0000527

PAHO WHO | Chikungunya. http://www.paho.org/hq/index.php?option=com_topics&view=article&id=343&Itemid=40931&lang=en. Accessed 29 Nov 2017

Petersen LR, Powers AM (2016) Chikungunya: epidemiology. F1000Research. https://doi.org/10.12688/f1000research.7171.1

Poo YS, Rudd PA, Gardner J, Wilson JAC, Larcher T, Colle M-A, Le TT, Nakaya HI, Warrilow D, Allcock R, Bielefeldt-Ohmann H, Schroder WA, Khromykh AA, Lopez JA, Suhrbier A (2014) Multiple immune factors are involved in controlling acute and chronic chikungunya virus infection. PLoS Negl Trop Dis 8:e3354. https://doi.org/10.1371/journal.pntd.0003354

Powers AM, Logue CH (2007) Changing patterns of chikungunya virus: re-emergence of a zoonotic arbovirus. J Gen Virol 88:2363–2377. https://doi.org/10.1099/vir.0.82858-0

Rahim AA, Thekkekara RJ, Bina T, Paul BJ (2016) Disability with persistent pain following an epidemic of chikungunya in rural South India. J Rheumatol 43:440–444. https://doi.org/10.3899/jrheum.141609

Ramachandran V, Kaur P, Kanagasabai K, Vadivoo S, Murhekar MV (2014) Persistent arthralgia among Chikungunya patients and associated risk factors in Chennai, South India. J Postgrad Med 60:3. https://doi.org/10.4103/0022-3859.128795

Ravi V (2006) Re-emergence of chikungunya virus in India. Indian J Med Microbiol, 83–84
Ravindran V, Alias G (2017) Efficacy of combination DMARD therapy vs. hydroxychloroquine monotherapy in chronic persistent chikungunya arthritis: a 24-week randomized controlled open label study. Clin Rheumatol 36:1335–1340. https://doi.org/10.1007/s10067-016-3429-0
Renault P, Solet J-L, Sissoko D, Balleydier E, Larrieu S, Filleul L, Lassalle C, Thiria J, Rachou E, de Valk H et al (2007) A major epidemic of chikungunya virus infection on Reunion Island, France, 2005–2006. Am J Trop Med Hyg 77:727–731
Robinson MC (1955) An epidemic of virus disease in Southern Province, Tanganyika territory, in 1952–1953. I. Clinical features. Trans R Soc Trop Med Hyg 49:28–32. https://doi.org/10.1016/0035-9203(55)90080-8
Rodriguez-Morales AJ, Restrepo-Posada VM, Acevedo-Escalante N, Rodríguez-Muñoz ED, Valencia-Marín M, Castrillón-Spitia JD, Londoño JJ, Bedoya-Rendón HD, Cárdenas-Pérez J de J, Cardona-Ospina JA, Lagos-Grisales GJ (2017) Impaired quality of life after chikungunya virus infection: a 12-month follow-up study of its chronic inflammatory rheumatism in La Virginia, Risaralda, Colombia. Rheumatol Int, 1–2. https://doi.org/10.1007/s00296-017-3795-1
Rodríguez-Morales AJ, Cardona-Ospina JA, Fernanda Urbano-Garzón S, Sebastian Hurtado-Zapata J (2016a) Prevalence of post-chikungunya infection chronic inflammatory arthritis: a systematic review and meta-analysis: chronic inflammatory rheumatism following chikungunya. Arthritis Care Res 68:1849–1858. https://doi.org/10.1002/acr.22900
Rodríguez-Morales AJ, Villamil-Gomez W, Merlano-Espinosa M, Simone-Kleber L (2016b) Post-chikungunya chronic arthralgia: a first retrospective follow-up study of 39 cases in Colombia. Clin Rheumatol 35:831–832. https://doi.org/10.1007/s10067-015-3041-8
Ross RW (1956) The Newala epidemic. III. The virus: Isolation, Pathogenic Properties and Relationship to the Epidemic. J Hyg (Lond) 54:177–191
Saluzzo JF, Cornet M, Digoutte JP (1983) Outbreak of a chikungunya virus epidemic in western Senegal in 1982. Dakar Med 28:497–500
Schilte C, Staikovsky F, Couderc T, Madec Y, Carpentier F, Kassab S, Albert ML, Lecuit M, Michault A (2013) Chikungunya virus-associated long-term arthralgia: a 36-month prospective longitudinal study. PLoS Negl Trop Dis 7:e2137. https://doi.org/10.1371/journal.pntd.0002137
Sergon K, Njuguna C, Kalani R, Ofula V, Onyango C, Konongoi LS, Bedno S, Burke H, Dumilla AM, Konde J, Njenga MK, Sang R, Breiman RF (2008) Seroprevalence of Chikungunya virus (CHIKV) infection on Lamu Island, Kenya, October 2004. Am J Trop Med Hyg 78:333–337
Sergon K, Yahaya AA, Brown J, Bedja SA, Mlindasse M, Agata N, Allaranger Y, Ball MD, Powers AM, Ofula V (2007) Seroprevalence of chikungunya virus infection on Grande Comore Island, union of the Comoros, 2005. Am J Trop Med Hyg 76:1189–1193
Shah KV, Gibbs CJ, Banerjee G (1964) Virological investigation of the epidemic of haemorrhagic fever in Calcutta: isolation of three strains of chikungunya virus. Indian J Med Res 52:676–683
Shegarfi H, Naddafi F, Mirshafiey A (2012) Natural killer cells and their role in rheumatoid arthritis: friend or foe? Sci World J 2012:1–10. https://doi.org/10.1100/2012/491974
Simon F, Parola P, Grandadam M, Fourcade S, Oliver M, Brouqui P, Hance P, Kraemer P, Mohamed AA, de Lamballerie X et al (2007) Chikungunya infection: an emerging rheumatism among travelers returned from Indian Ocean islands. Report of 47 cases. Medicine (Baltimore) 86:123–137
Simon F, Javelle E, Cabie A, Bouquillard E, Troisgros O, Gentile G, Leparc-Goffart I, Hoen B, Gandjbakhch F, Rene-Corail P, Franco J-M, Caumes E, Combe B, Poiraudeau S, Gane-Troplent F, Djossou F, Schaerverbeke T, Criquet-Hayot A, Carrere P, Malvy D, Gaillard P, Wendling D (2015) French guidelines for the management of chikungunya (acute and persistent presentations). November 2014. Médecine Mal Infect 45:243–263. https://doi.org/10.1016/j.medmal.2015.05.007
Sissoko D, Malvy D, Giry C, Delmas G, Paquet C, Gabrie P, Pettinelli F, Sanquer MA, Pierre V (2008) Outbreak of chikungunya fever in Mayotte, Comoros archipelago, 2005–2006. Trans R Soc Trop Med Hyg 102:780–786. https://doi.org/10.1016/j.trstmh.2008.02.018

Sissoko D, Malvy D, Ezzedine K, Renault P, Moscetti F, Ledrans M, Pierre V (2009) Post-epidemic chikungunya disease on Reunion Island: course of Rheumatic manifestations and associated factors over a 15-month period. PLoS Negl Trop Dis 3:e389. https://doi.org/10.1371/journal.pntd.0000389

Staikowsky F, Roux KL, Schuffenecker I, Laurent P, Grivard P, Develay A, Michault A (2008) Retrospective survey of chikungunya disease in Réunion Island hospital staff. Epidemiol Amp Infect 136:196–206. https://doi.org/10.1017/S0950268807008424

Teo T-H, Lum F-M, Claser C, Lulla V, Lulla A, Merits A, Renia L, Ng LFP (2013) A Pathogenic Role for CD4+ T Cells during chikungunya virus infection in mice. J Immunol 190:259–269. https://doi.org/10.4049/jimmunol.1202177

Thanapati S, Ganu M, Giri P, Kulkarni S, Sharma M, Babar P, Ganu A, Tripathy AS (2017) Impaired NK cell functionality and increased TNF-α production as biomarkers of chronic chikungunya arthritis and rheumatoid arthritis. Hum Immunol 78:370–374. https://doi.org/10.1016/j.humimm.2017.02.006

Thiberville S-D, Moyen N, Dupuis-Maguiraga L, Nougairede A, Gould EA, Roques P, de Lamballerie X (2013) Chikungunya fever: epidemiology, clinical syndrome, pathogenesis and therapy. Antiviral Res 99:345–370. https://doi.org/10.1016/j.antiviral.2013.06.009

Tomori O, Fagbami A, Fabiyi A (1975) The 1974 epidemic of chikungunya fever in children in Ibadan. Trop Geogr Med 27:413–417

Uhrlaub JL, Pulko V, DeFilippis VR, Broeckel R, Streblow DN, Coleman GD, Park BS, Lindo JF, Vickers I, Anzinger JJ, Nikolich-Žugich J (2016) Dysregulated TGF-β production underlies the age-related vulnerability to chikungunya virus. PLoS Pathog 12:e1005891. https://doi.org/10.1371/journal.ppat.1005891

van Aalst M, Nelen CM, Goorhuis A, Stijnis C, Grobusch MP (2017) Long-term sequelae of chikungunya virus disease: a systematic review. Travel Med Infect Dis 15:8–22. https://doi.org/10.1016/j.tmaid.2017.01.004

Venugopalan A, Ghorpade RP, Chopra A (2014) Cytokines in acute chikungunya. PLOS ONE 9: e111305. https://doi.org/10.1371/journal.pone.0111305

Wauquier N, Becquart P, Nkoghe D, Padilla C, Ndjoyi-Mbiguino A, Leroy EM (2011) The acute phase of chikungunya virus infection in humans is associated with strong innate immunity and T CD8 cell activation. J Infect Dis 204:115–123. https://doi.org/10.1093/infdis/jiq006

Waymouth HE, Zoutman DE, Towheed TE (2013) Chikungunya-related arthritis: case report and review of the literature. Semin Arthritis Rheum 43:273–278. https://doi.org/10.1016/j.semarthrit.2013.03.003

Win MK, Chow A, Dimatatac F, Go CJ, Leo YS (2010) Chikungunya fever in Singapore: acute clinical and laboratory features, and factors associated with persistent arthralgia. J Clin Virol 49:111–114. https://doi.org/10.1016/j.jcv.2010.07.004

Yaseen HM, Simon F, Deparis X, Marimoutou C (2014) Identification of initial severity determinants to predict arthritis after chikungunya infection in a cohort of French gendarmes. BMC Musculoskelet Disord 15:249. https://doi.org/10.1186/1471-2474-15-249

Zim MA, Sam I-C, Omar SFS, Chan YF, AbuBakar S, Kamarulzaman A (2013) Chikungunya infection in Malaysia: comparison with dengue infection in adults and predictors of persistent arthralgia. J Clin Virol 56:141–145. https://doi.org/10.1016/j.jcv.2012.10.019

Chikungunya Virus Vaccines: Platforms, Progress, and Challenges

Victor R. DeFilippis

Contents

Abstract Chikungunya is a clinically and economically important arbovirus that has spread globally in the twenty-first century. While uncommonly fatal, infection with the virus can lead to incapacitating arthralgia that can persist for months to years. The adverse impacts of viral spread are most severe in developing low- and middle-income countries in which medical infrastructure is insufficient and manual labor is an economic driver. Unfortunately, no prophylactic or therapeutic treatments are approved for human use to combat the virus. Historically, vaccination has proven to be the most efficient and successful strategy for protecting populations and eradicating infectious disease. A large and diverse range of promising vaccination approaches for use against Chikungunya has emerged in recent years and been shown to safely elicit protective immune responses in animal models and humans. Importantly, many of these are based on technologies that have been clinically approved for use against other pathogens. Furthermore, clinical trials are currently ongoing for a subset of these. The purpose of this review is to provide a description of the relevant immunobiology of Chikungunya infection, to present

V. R. DeFilippis (✉)
Vaccine and Gene Therapy Institute, Oregon Health and Science University,
505 NW 185th Ave, Portland, Beaverton, OR 97006, USA
e-mail: defilipp@ohsu.edu

Current Topics in Microbiology and Immunology (2022) 435: 81–106
https://doi.org/10.1007/82_2019_175

Published online: 24 July 2019

immune-stimulating technologies that have been successfully employed to protect against infection, and discuss priorities and challenges regarding the future development of a vaccine for clinical use.

1 Introduction: Chikungunya Ecology and Economic Impact

Chikungunya virus (CHIKV) is an arthritogenic arbovirus belonging to the Old World clade of the alphavirus genus *Togaviridae*. As with other clinically impactful arboviruses, acute infection with CHIKV causes fever and severe joint pain. However, while CHIKV is not typically fatal it may lead to chronic and long-lasting arthralgic sequelae that can be debilitating (1, 2). Interhost transmission of CHIKV occurs via the bite of *Aedes* (subgenus *Stegomyia*) mosquito species. In regions where CHIKV is endemic, including tropical and subtropical areas of sub-Saharan Africa and Southeast Asia, two separate maintenance and transmission modes are evident. Enzootic cycles within nonhuman primate (NHP) species such as African green monkeys (*Cercopithecus aethiops*) and baboons (*Papio ursinus*) are believed to represent the evolutionarily ancestral maintenance pattern in forest and savannah (Tsetsarkin et al. 2016). Sporadic spillover of the virus into urban areas is thus likely to be the primary source of human CHIKV outbreaks (Diallo et al. 2012). While the sustained human–mosquito–human transmission is largely attributed to *Aedes aegypti* or *Aedes albopictus* species, the crucial factors that drive emergence into urban venues are not known. Nonetheless, the reality of ongoing enzootic transmission and periodic but unpredictable urban spillover means that in the absence of vector elimination strategies, human outbreaks will occur with regularity. Importantly, this also makes eradication of the virus by way of clinical, human host-administered treatments improbable. However, human populations could be rendered resistant to infection and thus limit spread through the generation of virus-specific immunity elicited by vaccination. While no CHIKV vaccines are currently approved for human use, many promising, safely efficacious, and novel vaccine platforms have been described including some that are currently in clinical trials.

While CHIKV-associated fatality rates are low, the economic impact of disease can be extreme and disproportionate between health systems and tiers of societal development. This includes the direct medical expenses associated with acute illness such as symptomatic treatment (de Brito et al. 2016) but also chronic inflammatory conditions which are more protracted and costly (Cardiel 2012). Perhaps more importantly, large indirect financial impacts result from acute and chronic CHIKV including illness-associated work absence, diminished productivity, and recurrent transportation to health providers. These indirect costs can be particularly high since many countries experiencing CHIKV outbreaks are underdeveloped and their economic capacities rely heavily on manual labor (Seyler et al. 2010). One analysis estimated that between 2013 and 2015 in the Americas >23 million disability-adjusted life years were lost leading to >$83 billion in direct medical costs and >$101 billion in indirect societal costs (Bloch 2017). In addition, regions susceptible to CHIKV transmission often represent destinations popular with tourists. In line with this, the 2005–2006 outbreak on the island of La Reunion led to an estimated loss of $160 million in this industry (Enserink 2007). An efficacious vaccine would thus confer economic benefits in multiple ways including preventing incapacitation of labor forces, diminishing direct medical costs, and mitigating tourist uncertainty.

2 CHIKV Replication and Immune Reactivity

The CHIKV genome is approximately 11.8 kb single-stranded positive sense RNA that includes 5' and 3' untranslated regions (UTR), a 3' poly-A tail and 5' 7-methylguanosine cap [reviewed in (Vaney et al. 2013)]. Two open reading frames (ORFs) encode polyproteins separately comprised of four nonstructural (nsP1-4) and six structural (capsid [C], 6 K, transframe [TF], envelope [E] 1, E2, E3) proteins. Infectious CHIKV exists as a 70 nm enveloped virion composed of 240 copies of E1 and E2 heterodimers in an icosahedral lattice configuration surrounding a nucleocapsid (Voss et al. 2010). E1 and E2 play essential molecular roles during initiation of replication that involve virion interaction with cellular receptors and internalization by clathrin-mediated endocytosis (Strauss and Strauss 1994). Following the release of the nucleocapsid into the cytoplasm, viral genomic RNA is translated to generate the nsP polyprotein, the proteolytic maturation of which gives rise to the RNA replicase [reviewed in (van Duijl-Richter et al. 2015)]. This is followed by the translation of the structural polyprotein from a subgenomic promoter at the terminus of the nsP coding region. Proteolysis of the structural proteins enables nucleocapsid interaction with genomic RNA, association with envelope glycoproteins at the plasma membrane, and budding of assembled particles.

Humoral and cellular adaptive immune processes are potently induced in response to CHIKV infection. Cell-mediated immunity is triggered and CD8+ and CD4+ T cells reactive to CHIKV antigens are known in humans (Hoarau et al. 2013), nonhuman primates (Messaoudi et al. 2013), and mice (Morrison et al. 2011; Teo et al. 2012a). The role for T cells in viral clearance, however, is not well understood and, in fact, may contribute to disease exacerbation (Teo et al. 2012a). In contrast, numerous studies have demonstrated the importance of antibodies with neutralization capacity for controlling CHIKV infection and tissue damage (Couderc et al. 2009; Sheela and Sumathy 2013; Lum et al. 2013; Hawman et al. 2013). IgM and IgG antibodies reactive to CHIKV appear early after symptom onset and IgG persists at high levels for years (Couderc et al. 2009; Kam et al. 2012a). Prior infection and development of anti-CHIKV neutralizing antibodies in humans associated with protection from re-infection (Yoon et al. 2015; Kam et al. 2012b; Galatas et al. 2016). In addition, passively transferred serum from convalescent humans protects mice from virus-associated lethality (Couderc et al. 2009). Intriguingly, CHIKV-reactive monoclonal antibodies have been isolated from patients that can also protect mice and even lead to diminished viremia and virus-associated disease in nonhuman primates (Smith et al. 2015; Broeckel et al. 2017). Longitudinal analysis in humans has identified the E2 and E3 glycoproteins as well as capsid and nsP3 as key targets of the antibody response (Kam et al. 2012a, c). However, the E2 protein appears to represent the immunodominant viral protein and antibodies reactive to it can confer broad-based viral neutralization (Kam et al. 2012c; Fox et al. 2015; Weger-Lucarelli et al. 2015). Surprisingly, no established minimum protective antibody titers have been described, although a reciprocal titer of approximately 35 was shown to be required in IFN-deficient mice, which presumably exhibit heightened susceptibility to the virus in comparison to humans (Chu et al. 2013).

Four main phylogenetic clades of CHIKV are recognized and include East/Central/South African (ECSA), West African, Asian, and Indian Ocean lineages (IOL) that collectively share 84.5–97.8% amino acid identity (Sahadeo et al. 2017; Volk et al. 2010). ECSA and Asian lineages are considered to be older and having evolved for transmission via *A. aegypti*. In recent years, however, adaptive mutations arose in the viral E1/E2 proteins of the IOL that likely facilitated interhost transmission by the more anthropophilic *A. albopictus,* a factor potentially playing a role in the virus' explosive global emergence (Tsetsarkin et al. 2007; Tsetsarkin and Weaver 2011). However, virus derived from the Asian lineage appears to be responsible for the emergence and local transmission in the Americas, which was first observed in 2013 (Lanciotti and Valadere 2014; Leparc-Goffart et al. 2014). While differences in virulence have been observed between viral lineages in mice (Wei Chiam et al. 2015; Teo et al. 2015) [especially those deficient in the type I interferon response with WA strains appearing more lethal (Langsjoen et al. 2018)] whether these or other differences are manifest in humans is not known. Interestingly, high CHIKV levels were detected in blood from healthy donors in Puerto Rico during an epidemic in 2014 that involved Asian/American lineages, suggesting that these strains may associate with higher rates of asymptomatic infection (Simmons et al. 2016). Importantly, from the standpoint of vaccination and control efforts, all CHIKV lineages comprise one serotype and cross-protection between strains is both observable and enduring in humans and other species (Langsjoen et al. 2018; Auerswald et al. 2018). Furthermore, antibodies reactive to CHIKV antigens may even confer protection against related Alphaviruses such as Ross River (Gardner et al. 2010) and O'nyong'nyong viruses (Partidos et al. 2012). However, despite this, variation has been reported in the neutralization efficiency of convalescent human sera across ECSA and Asian clades (Chua et al. 2016) suggesting that evaluation of such issues in vaccination efficacy studies is worthwhile. Nevertheless, vaccines that are, in general, antigenically focused are likely to protect against circulating lineages from all clinically relevant clades.

3 Animal Models of CHIKV Infection and Immunity

Fortunately for the purposes of vaccine and therapeutic development, tractable animal models that faithfully recapitulate many aspects of human infection are available for examination of CHIKV replication, immunity, and pathogenesis [reviewed in (Haese et al. 2016)]. These have been instrumental for both investigating the safety and efficacy of novel vaccination platforms as well as characterizing the associated immune-mediated mechanisms. The most widely employed are wild-type and type I interferon (IFN) deficient mouse models (Taylor et al. 2015; Teo et al. 2012b). As experimental systems, these exhibit many desirable attributes including relatively low cost, ease of handling and care, ability to control environmental and genetic variables, and availability of reagents. They can also be utilized for studies of acute, chronic, and lethal CHIKV infection as well as virus-associated inflammatory dysfunction (Morrison et al. 2011; Hawman et al. 2013; Ziegler et al. 2008; Couderc et al. 2008a; Poo et al. 2014). In particular, lack of type I IFN-dependent antiviral phenotypes renders animals fully susceptible to CHIKV-induced mortality (Couderc et al. 2008b; Rudd et al. 2012; Gardner et al. 2012), and, as such, this represents a highly stringent model for exploring the potency of antibodies and T cells generated by vaccination or following inter-animal transfer (Smith et al. 2015; Plante et al. 2011; Pal et al. 2013). Furthermore, many fundamental innate and adaptive immunologic processes involved in CHIKV protection and clearance are shared between humans and mice. As such they represent a powerful and cost-effective first-pass testing platform for characterization and mechanistic understanding of novel vaccine technologies. Informative nonhuman primate (NHP) models of CHIKV infection and immunity have also been developed and are commonly used, including Cynomolgus (Chen et al. 2010; Messaoudi et al. 2013) and Rhesus (Chen et al. 2010; Messaoudi et al. 2013) macaque species. NHP models exhibit advantages over murine models that make them more translatable to humans such as increased susceptibility to infection, outbred genetic structure, and maternal/fetal transmission. In addition, NHP models are considered more representative of human vaccine responses owing to their size, distribution of immune cell subsets, innate immune receptor expression, and the morphology of injection site and lymphatic systems [reviewed in (Thompson and Loré 2017)]. However, their expense and requirement for highly specialized housing render NHP models more applicable to studies examining vaccine technologies for which a more comprehensive proof of principle has been established.

4 Vaccine Technology Classes

Given the importance of humoral immunity for controlling infection, ongoing vaccine development efforts are primarily focused on eliciting antibody-dependent responses for protection against CHIKV. To accomplish this, five general technological platforms have been or are being explored. These include live attenuated chikungunya viral (LAV) strains, chimeric viral vectors, inactivated virions or virus-like particles (VLP), subunit vaccines, and DNA plasmid. Each has displayed efficacy in animal models and some are even immunogenic in humans. A LAV for CHIKV involves limited intrahost replication and has proven particularly capable of eliciting neutralizing antibody production in humans (Edelman et al. 2000). However, as a class, such vaccines are often associated with suboptimal safety profiles owing mostly to their unlikely but demonstrated the potential for reversion to a pathogenic viral state. As such they are contraindicated in hosts with weakened or immature immune systems including the elderly, neonates, and those undergoing strategic immunosuppression. This has necessitated the development of novel immunogen delivery platforms that do not involve viral growth within vaccinees. Moreover, while direct administration of inert viral antigen or molecular platforms that involve the synthesis of such antigen in vivo are safer, these often lead to weaker, shorter lived immunity since they stimulate fewer facets of the immune response in a localized, rather than disseminated manner. As such they often require co-delivery of nonantigenic molecules termed adjuvants that act by stimulating localized innate immune processes to enhance immunogenicity [reviewed in (Reed et al. 2013; Coffman et al. 2010)]. Alternatively, recombinant heterologous viral platforms that encode and express CHIKV antigen but are incapable of growth in vivo represent another safe and powerful method for eliciting protective immunity [reviewed in (Ramsauer and Tangy 2016)]. Diverse efforts pursuing these and other technologies are currently the subject of substantial research investment to identify an optimal method of immune protection against CHIKV.

4.1 Live Attenuated Viral Vaccines

The earliest CHIKV vaccine to successfully demonstrate protective immunostimulation in humans involved a clinical strain from Thailand that was sequentially passaged multiple times on human MRC5 human lung fibroblasts and designated 181/clone 25 (Levitt et al. 1986). In clinical trials, the vaccine was highly immunogenic with all but one out of 59 adults generating neutralizing antibody titers (Edelman et al. 2000). However, 10% of recipients experienced mild arthralgia. It was subsequently determined that only two amino acid changes were responsible for the pathogenic differences between the clinical and the attenuated vaccine strain, making the likelihood of reversion unacceptably high for clinical use (Gorchakov et al. 2012). Other efforts to construct LAV have involved modifying targeted genomic regions to impair viral growth. This is greatly facilitated by utilizing infectious cDNA clones for viral reconstitution, the mutation of which is straightforward by standard molecular methods (Vanlandingham et al. 2005; Tsetsarkin et al. 2006). For instance, CHIKV-containing deletions in open reading frames that encode nsP3, 6 K, or E2 have each been shown to be replication incompetent in vivo yet capable of eliciting protective antibody responses in mice (Hallengärd et al. 2014; Piper et al. 2013; Gardner et al. 2014). The most promising implementation of this approach is the Δ5nsP3-attenuated virus from the laboratory of Peter Liljestrom that contains a 60 amino acid deletion of the nsP3 protein (Hallengärd et al. 2014). This attenuation strategy allows for sufficient in vitro replication but no viremia after inoculation of mice with a dose sufficient to elicit neutralizing antibody titers and antiviral T cell responses and block detectable viremia. This vaccine has proved to be similarly efficacious in a cynomolgus macaque viral immunogenicity and challenge model (Roques et al. 2017). Importantly, a phase I clinical trial using this platform (identifier NCT03382964) that involved 120 participants was very recently completed by Valneva and showed a superior safety profile and seroconversion in all vaccinees at day 14 after a single dose.

A rational attenuation method devised by Scott Weaver and colleagues involves replacing the endogenous subgenomic viral promoter region responsible for driving normal expression of antigenic structural proteins with a picornavirus internal ribosomal entry site (IRES), leading to diminished protein synthesis (Chu et al. 2013; Langsjoen et al. 2018; Partidos et al. 2012; Plante et al. 2011, 2015; Roy et al. 2014; Kim et al. 2011). The resulting virus grows slowly in vitro, is fully attenuated in vivo, and does not replicate in mosquito cells so it cannot be conventionally transmitted between hosts. Importantly, given the size and complexity of the modified genomic region, reversion mutations are quite improbable making it a presumably safe treatment. This vaccine effectively generates protective antibody responses in both wild-type and IFN-deficient mice as well as NHP without adverse effects. While conventional T cell activation appears to be triggered by the vaccine, adoptive transfer of CD4+ or CD8+ cells to mice following vaccination did not protect against challenge; an additional indication that cellular immunity is of limited relevance during CHIKV infection. Moreover, the vaccine's efficacy in immunocompromised animals further predicts its safety for immune-insufficient patients. This vaccine was also found to elicit protective immunity against the antigenically related Alphavirus O'nyong'nyong virus (ONNV) in IFN-deficient mice and dams receiving the vaccine vertically transferred antibody-mediated protection to their offspring. This platform is currently being developed commercially by a partnership between Takeda Pharmaceuticals and Zydus Cadila.

4.2 Chimeric Viral Vector Platforms

Another impactful vaccination approach being actively explored involves the delivery of CHIKV antigen using heterologous viral vectors [reviewed in (Ramsauer and Tangy 2016)]. This includes the construction of chimeric virus particles containing CHIKV genomic regions encoding antigenic proteins that enter cells in vivo but either do not generate progeny or only undergo a single round of replication. Their limited viability renders them unlikely to confer pathogenic or reactogenic effects. Virus types co-opted for this purpose include both RNA- and DNA-based agents, including related Alphaviruses. Moreover, depending on the platform they are also able to elicit both humoral and cell-mediated immune responses especially since, by utilizing cellular protein synthesis and processing pathways, they lead to antigen presentation via MHC class I and subsequent T cell activation (Liu 2010). Being structurally complete viruses, these vectors are also likely to be comprised of molecular patterns that activate pattern recognition receptors and thus enhance immunogenicity through innate immune stimulation (Hu and Shu 2018; Iwasaki and Medzhitov 2015). While preexisting immunity against some common human viruses can lead to diminished efficacy of vectors on which they are based (Saxena et al. 2013), solutions appear to be obtainable including utilizing nonhuman virus types or antigenically modifying human viruses. As with LAV, chimeric viruses must be maintained and administered in a manner that permits viability as determined by cellular fusion and genome delivery. As such, both technologies can require specialized manufacturing, storage, and transport procedures that are not always feasible with application in underdeveloped areas (Ulmer et al. 2006; Lloyd and Cheyne 2017).

The earliest chimeric vaccines against CHIKV involved vectors that encode the nonstructural proteins of other related alphavirus species including Sindbis virus (SINV), Eastern equine encephalitis virus (EEEV), or an attenuated strain (TC-83) of Venezuelan equine encephalitis virus (VEEV) with the structural proteins of CHIKV (Wang et al. 2008, 2011a). This allowed generation of virus particles that grow to high titer in cell culture but do not replicate in vivo, indicative of appropriate safety. Intriguingly, the VEEV- and EEEV-based vectors also elicited neutralizing antibody titers in vaccinated mice and led to protection from viremia and virus-associated mortality and disease in these animals. VEEV-based vectors were additionally modified by the introduction of IRES elements to drive structural protein synthesis, further improving their safety, rendering them effective in IFN-deficient mice, and preventing their mosquito transmissibility.

Vesicular stomatitis virus (VSV) represents another RNA-based virus that can be engineered using simple methods and exploited for heterologous vaccination purposes (Roberts et al. 1999). VSV is a negative-stranded RNA member of the Rhabdoviridae family whose envelope can be pseudotyped with glycoproteins derived from unrelated viruses to enhance immunogenicity and alter cellular tropism. In addition, VSV-based vectors are able to elicit potent CD8+ T cell responses against target antigens (Haglund et al. 2002) and have been successfully employed in the context of diverse viral and bacterial pathogens. Chattopadhyay and colleagues have constructed chimeric VSV in which the endogenous glycoprotein G coding region is replaced by CHIKV structural proteins (Chattopadhyay et al. 2012). This yields virions into which antigenic CHIKV envelope glycoproteins are incorporated but that also enables cellular entry and new intracellular synthesis of these antigens following inoculation. The vector was shown in mice to elicit both neutralizing antibody and cellular immune responses against CHIKV and protection against viremia and CHIKV-associated disease. Whether CHIKV-directed vaccines that incorporate this platform operate safely in NHP models or humans still requires examination but currently a VSV-based vaccine against Ebola virus is undergoing phase III clinical trials (Coller et al. 2017).

One particularly novel chimeric virus vaccine was recently described by Erasmus and colleagues that involves Eilat virus (EILV), an alphavirus specific to insect species (Erasmus et al. 2015, 2016). The virus is incapable of replication in mammals due to an inability to enter cells or form nonstructural protein-mediated replication complexes (Nasar et al. 2015). Based on this, it was predicted that chimeric genomes consisting of the EILV 5'UTR and nonstructural polyprotein and CHIKV structural polyprotein would yield antigenic virus particles when introduced into mosquito cells. Furthermore, despite containing intact genomes and CHIKV virion proteins, the virus should not be capable of generating a replication competent, and thus pathogenic, particles in vivo since the nsPs do not function in mammalian cells. Moreover, the lack of viral RNA transcription also precludes the emergence of replication-viable mutants. Remarkably, a single dose of the vaccine was shown to elicit neutralizing antibodies and protection against viral challenge in WT and IFN-deficient mice as well as NHP. While the vaccine's efficacy has not been comparatively evaluated relative to other viral-vectored or VLP-based platforms, it may involve an enhanced mechanism of antigen presentation by entering cells via the endocytic pathway (Erasmus et al. 2016). In addition, the presence of genomic RNA in the viral particle may provoke innate immune reactivity and thus adjuvant-like properties (Vasilakos and Tomai 2013). As such, EILV may represent a powerful vaccine platform for CHIKV and other viruses that warrants thorough exploration.

DNA-based viruses have also been shown to be effective as vectors for vaccine antigen delivery and immunostimulation. This includes adenoviruses and the poxvirus modified vaccinia virus Ankara (MVA) (Volz and Sutter 2017; Majhen et al. 2014). Non-replicating adenovirus vectors have been well-studied for use in both gene therapy and vaccination and this includes hundreds of clinical trials. As such, ample evidence of their safety and efficacy has been amassed. The complex adenovirus 5 (Ad5) vaccine vectors contain deletions of regions essential for replication that can be complemented in transgenic cell lines to allow particle formation. These are amenable to accepting large user-defined insertions and have been effectively employed as vaccines against viral pathogens including influenza, Ebola, Dengue, West Nile, and Rift Valley fever viruses. Wang and colleagues used this platform to deliver CHIKV envelope glycoproteins and capsid to CD-1 and C57BL/6 mice (Wang et al. 2011b). They show that the vaccine elicits antibody titers as high as those generated by natural infection and confers complete protection against viral challenge. Additionally, adenoviruses of nonhuman species also represent potentially safe and efficacious vaccine vectors that circumvent any impacts of preexisting immunity in recipients. This is currently being employed in clinical trials using Chimpanzee adenovirus expressing CHIKV structural proteins (identifier NCT03590392) (Ewer et al. 2017).

MVA is derived from vaccinia virus smallpox vaccine following 570 passages in avian tissue culture, which caused large genomic deletions that abrogate productive replication in many species including humans (Moss et al. 1996). MVA also supports stable introduction and expression of heterologous antigenic proteins for which it has been effectively utilized to stimulate immunity against viral, bacterial, and parasitic pathogens. Importantly, MVA also leads to the generation of both humoral and cell-mediated protective immunity (Draper et al. 2013). In 2014, three separate groups reported the synthesis and assessment of MVA-vectored vaccines against CHIKV. Van den Doel and colleagues constructed a vector that expresses CHIKV 6K-E1-E3-E2 polyprotein and fully protects IFN-deficient A129 mice from CHIKV-induced lethality without adverse effects (van den Doel et al. 2014). Weger-Lucarelli and colleagues similarly generated MVA that expresses only E2E3 (Weger-Lucarelli et al. 2014). This also protected A129 mice from lethal CHIKV challenge. Interestingly, however, sera from mice vaccinated with this vector failed to protect nonvaccinated A129 mice from CHIKV mortality following the passive transfer. Moreover, depletion of CD4+ , but not CD8+ cells from vaccinated mice prevented MVA-mediated protection, suggestive of a role for cell-mediated immunity in this outcome. Finally, Garcia-Arriaza and colleagues inserted the entire CHIKV structural polyprotein (C-E3-E2-6K-E1) into MVA and demonstrated strong innate immune induction following the vector's administration, consistent with that observed during conventional viral infection or injection of adjuvanted vaccines (discussed below) (García-Arriaza et al. 2014). This vector clearly elicited a broad, polyfunctional T cell response as well as neutralizing antibodies and protection against viremia in C57BL/6 mice. Currently, both adenovirus and MVA-based CHIKV vaccines are in a preclinical stage of development and NHP studies would greatly enhance our evaluation of their translatability to humans. Importantly, however, these platforms also represent important research tools that can lead to useful mechanistic insight regarding the respective roles of humoral and cell-mediated immune processes for protection against CHIKV and the development of CHIKV-associated disease.

The most clinically advanced chimeric platform for CHIKV immunization involves the Measles virus (MV) Schwarz attenuated vaccine strain as a delivery vector [reviewed in (Tangy and Naim 2005)]. MV is a negative-stranded RNA virus for which a cDNA plasmid-based reverse genetic system has been optimized that allows expression of heterologous protein antigens (Combredet et al. 2003). MV/Schwarz is an extremely safe, efficacious, and historically utilized vaccine that confers lifelong protective immunity following a single immunization. Furthermore, it has been effectively adapted for the generation of protective responses against viral pathogens such as West Nile, dengue, HIV, and SARS coronavirus [reviewed in (Mühlebach 2017)]. For purposes of practical investigation and clinical translatability, the MV platform exhibits important advantages. First, transgenic mouse models have been constructed to be susceptible to MV infection through the introduction of the viral receptor CD46 in the presence and absence of a functional IFN response for added stringency (Combredet et al. 2003). Next, NHP are also infectible with MV and immune reactivity to encoded antigens occurs in the following exposure (de Swart and DeSwart 2009). Finally, preexisting immunity to MV in vaccines does not appear to impact heterologous immunogenicity elicited by the vector in animals or humans (Reisinger et al. 2019).

Frederic Tangy's research group has elegantly leveraged this technology to synthesize a safe and immunogenic CHIKV vaccine (Brandler et al. 2013;

Ramsauer et al. 2015). Specifically, MV/Schwarz genomes have been constructed that contain the full CHIKV structural polyprotein that replicates to high titer in vitro and confers potent antibody-mediated protective immunity against multiple clades of CHIKV in IFN-deficient mice as well as cynomolgus macaques (Rossi et al. 2019). Importantly, this vaccine has also completed phase II clinical trials (identifiers NCT03101111, NCT02861586) clinical trials to demonstrate safety, tolerability, and immunogenicity (Reisinger et al. 2019). In the first trial, 36 subjects received one of three MV-CHIKV doses and were boosted after either 1 or 3 months. Single immunization led to detectable neutralizing antibody titers for all doses. However, full seroconversion was achieved in all vaccine groups after the booster immunization, preexisting anti-MV antibodies did not affect the immune response, and the vaccine displayed acceptable tolerability. The second trial involved 236 participants that received control MV (n = 34), MV-CHIK (n = 195), or MV control and MV-CHIK (n = 34). This study demonstrated seroconversion of 50–93% of participants after a single MV-CHIK dose and 86–100% of participants after MV-CHIK boost. Moreover, antibody titers were positively correlated with vaccine dose and preexisting immunity to MV did not affect reactivity to MV-CHIK. It is important to note that the precise MV-CHIKV mechanism of action is not defined. However, cells infected in vitro with the virus are known to synthesize and release virus-like particles (VLP) that are likely to be immunogenic (Brandler et al. 2013) (discussed in more detail below). It is possible that these are taken up and trafficked to lymph nodes by antigen presenting cells to initiate immune responses. Currently, this vaccine is being commercially developed by Themis Bioscience.

4.3 Inert Antigen Vaccines

In contrast to vaccine platforms that involve molecular synthetic processes during virus-based antigen delivery, others exist that are comprised of inactivated or genome-deficient viral particles or individual purified viral proteins. Given the spectrum of immunocompetence present in the human population, these are generally considered safer than LAV and chimeric virus vectors since they do not undergo any in vivo molecular replication. However, the antigenicity of these may be suboptimal since conformation of native epitopes can be susceptible to damage in response to the inactivation and purification processes. In addition, these are administered locally and without regional amplification or systemic immune activation. As such, inert virus particle-based vaccines exhibit lower immunogenicity and can require booster immunizations as well as the inclusion of adjuvants to stimulate innate immune processes and enhance antigen processing and delivery. Multiple methods have been utilized to inactivate live CHIKV particles in a manner that aims to maintain their immunogenic capacity. This includes treatment with chemicals such as binary ethyleneimine (Gardner et al. 2010), β-propiolactone (Kumar et al. 2012), Tween 80/ether (Eckels et al. 1970), and formalin

(Tiwari et al. 2009) or exposure to UV-irradiation (Nakao and Hotta 1973). These have been examined in preclinical mouse or NHP models and proved to generate protective virus-directed antibody responses, in some cases even without adjuvant. Formalin-inactivated CHIKV has also been used in human volunteers without adjuvants and shown to tolerably elicit antibodies with neutralization capacity (Harrison et al. 1971). Unfortunately, inactivation of live clinical strains of CHIKV requires biosafety level 3 conditions for production. This renders the process costly and difficult to scale-up and thus only feasible for experimental work.

Antigenic fidelity of CHIKV proteins for efficacious vaccine-mediated immunostimulation can also be readily achieved through synthesis and purification of virus-like particles (VLP). These are comprised of self-assembling viral structural proteins that conformationally resemble wild-type virions but contain no genomic material and thus cannot initiate replication [reviewed in (Mohsen et al. 2017)]. The safety profile of VLPs is thus heightened over inactivated, live attenuated, or chimeric viruses since viral growth is not possible in any step of the production or vaccination process. In addition, biosafety requirements for manufacturing are also less stringent and thus scale-up is more technically feasible and cost-effective. Generation of virion-based antigens on mammalian or insect (Metz et al. 2013a) cells additionally benefits from proper protein folding, glycosylation, and post-translational modifications important for epitope structure. Currently, several licensed VLP-based vaccines are in widespread clinical use against human papillomavirus, hepatitis B, and hepatitis E viruses. In the case of CHIKV, ectopic translation of the viral envelope polyprotein (E2-E2-6K-E1) in cultured cell lines following transfection or transduction of expression vectors leads to the secretion of particles that are similar in size, structure, and protein configuration to the native virus (Akahata et al. 2010; Sun et al. 2013). These can then be isolated, purified, and concentrated from cell culture media by ultracentrifugation or filtration for direct in vivo administration. Importantly, human cell lines amenable to CHIKV VLP synthesis have been approved for vaccine production by regulatory agencies and thus tractable methods of clinic-ready scale-up are available (Genzel 2015).

Studies have demonstrated that immunization with VLP synthesized using mammalian or insect cells elicit protective antibody responses in wild-type and an IFN-deficient mouse models as well as in NHP (Metz et al. 2013a, b; Akahata et al. 2010). Importantly, a VLP-based vaccine manufactured using characterized and FDA-approved HEK293 cells by the NIH Vaccine Research Center has completed a phase I clinical trial (identifier NCT01489358) (Chang et al. 2014) and phase II trials in the Caribbean (identifier NCT03483961, NCT02562482) are currently ongoing. The completed phase I study involved 23 participants receiving one of three doses at 0, 4, and 20 days by intramuscular injection. The vaccine was well tolerated and, crucially, led to the production of neutralizing antibodies after the first boost in a manner weakly associated with antigen dose for all participants. Titers were maintained for at least 6 months and were at levels likely to be protective based on observations from convalescent sera (Kam et al. 2012a, b). Moreover, the neutralization capacity of sera collected from participants was shown to exist against nine viral strains comprising three clades, suggesting that vaccination against one strain can elicit cross-protection

(Goo et al. 2016). Interestingly, while published animal studies have employed CHIK VLP admixed with innate stimulating adjuvants, the clinical trial actually observed neutralizing antibodies in their absence. Whether adjuvants are not necessary or, rather, could actually enhance protective immune responses to a degree that allows dose sparing or single administration will necessitate further exploration. The duration that protective antibody titers are measurable following vaccination also requires investigation and this may also warrant the inclusion of adjuvants.

Individual recombinant CHIKV envelope proteins have also been effectively utilized as a vaccine antigen. Immunization with single proteins offers advantages such as ease and simplicity of synthesis and adaptability to strain variation if needed. Bacterial expression systems were used to synthesize and purify E1 and E2 proteins (Kumar et al. 2012; Khan et al. 2012). Administration of these proteins to WT mice elicited neutralizing antibodies in WT mice as well as protection from viremia, viral dissemination, and virus-mediated disease. However, recombinant peptides alone appear to be poorly immunogenic and require adjuvant admixture. Multiple classes of adjuvants were shown to enhance the E1- and E2-associated immune response. In addition, vaccination with E1/E2 subunits is also capable of generating balanced Th1/Th2 polarization as indicated by high titers of IgG1, IgG2c, and IgG3 antibody subclasses. The efficacy of subunit vaccination has yet to be examined in IFN-deficient mice or non-murine models.

4.4 Nucleic Acid Based Vaccines

Another approach to CHIKV vaccination involves the utilization of plasmid DNA that encodes immunogenic or protective factors that prevent infection. Direct delivery, often involving intracellular electroporation, into receptive tissues of DNA vectors can lead to in situ synthesis of user-specified proteins. This technology exhibits many logistical and biological advantages including stimulation of both humoral and cell-mediated immune responses, low production cost, high vaccine stability (cold chain independence), ease of manufacturing, capacity for co-delivery of multiple antigens, and capacity for rapid customization in the event of spontaneous strain emergence. Moreover, only basic molecular cloning methods are needed for synthesizing and adapting vectors of interest, greatly facilitating the experimental investigation. Muthumani and colleagues described the first DNA-based vaccine vector for CHIKV (Muthumani et al. 2008). The authors designed consensus E1, E2, and capsid proteins based on multiple sequences derived from clinical isolates across years of outbreaks and cloned the respective coding regions into a DNA expression vector. Intramuscular electroporation was used to deliver the constructs to mice and shown to elicit both CHIKV-reactive antibodies and IFNγ-producing T cells in response to CHIKV antigen. Follow up work by this group included the construction of an E1, E2, and E3 expressing vector that led to similar immune induction as well as protection against viral challenge in mice (Mallilankaraman et al. 2011). Additionally, this vector was also able to induce neutralizing antibody and T cell responses following administration to NHP indicating that it is operational across species. A similar approach that has also been used involves the introduction to DNA vectors of CHIKV-derived genomic regions that allow the in vivo synthesis and secretion of immunogenic VLP (Hallengärd et al. 2014; Hallengard et al. 2014) or the LAV strain 181/25 (Tretyakova et al. 2014). Both methods led to the production of neutralizing antibody responses in mice and while protection against CHIKV challenge was not checked for the LAV, it was observed for the VLP as indicated by the absence of viremia and CHIKV-associated disease. An antigen-independent DNA vector-based approach has also been used to successfully elicit prophylactic immunity to CHIKV infection in a manner that is more rapid than traditional vaccination. This involves the DNA encoding of an envelope-reactive monoclonal antibody sequence with broadly neutralizing capacity (Muthumani et al. 2016). In vivo delivery of this vector led to the synthesis of detectable CHIKV-reactive antibodies after three days and peak titers were observed by day 15–21. Remarkably, this vector conferred protection against CHIKV when administered as early as two or as late as thirty days pre-challenge. While DNA-based vaccines hold much promise, additional mechanistic and efficacy studies will be needed to fully elucidate their potential clinical applicability.

In addition to antigenic CHIKV proteins being encoded by DNA, they may also be expressed in vivo directly from messenger RNA (mRNA). The susceptibility of RNA to degradation leads to more stringent delivery requirements that protect the molecule such as lipid nanoparticle formulations. This approach has been used to deliver liposomal mRNA that encodes a highly potent neutralizing antibody identified from a B cell clone of a convalescent patient. This has led to vaccine dose-associated protection of IFN-deficient mice from lethal CHIKV challenge (Kose et al. 2019). Furthermore, administration of this platform in WT mice also led to substantial reduction or prevention of CHIKV viremia and viral RNA tissue loads and its use in cynomolgus macaques led to detectable serum levels of encoded IgG. Finally, mRNA encoding CHIKV antigens has proven capable of eliciting neutralizing antibodies in human subjects (Shaw et al. 2019). A phase I clinical trial using this vaccine is currently ongoing under the direction of Moderna, Inc. (identifier NCT03325075).

5 Priorities for and Barriers to Clinical CHIKV Vaccine Development

As outlined above and summarized in Table 1, numerous vaccine technologies have been shown to safely and efficaciously elicit protection in animal and human models against CHIKV infection and disease. A number of these platforms have even obtained regulatory approval for use against other pathogens. In addition,

Table 1 Vaccine list

Platform	Name	Description	Models	Development	References
LAV	181/25 (TSI-GSD-218)	Clinical strain attenuated by serial passage	Mouse, NHP, Human	Phase II (halted)	Edelman et al. (2000), Levitt et al. (1986)
LAV	LR-E2-E79 K	Clinical strain LR attenuated by mutations in heparin sulfate binding domains	WT, IFN-deficient mice	Preclinical	Gardner et al. (2014)
LAV	Δ5nsP3, Δ6 K	Clinical strain attenuated by targeted mutations in nsP3, 6 K	WT Mice, NHP	Phase I	Hallengärd et al. (2014), Roques et al. (2017)
LAV	CHIKV TM17-2	Clinical strain attenuated by mutations in E2 transmembrane domain	WT Mice	Preclinical	Piper et al. (2013)
LAV	CHIKV/IRES	Subgenomic promoter replaced with IRES	WT, IFN-deficient mice, NHP	Preclinical	Chu et al. (2013), Langsjoen et al. (2018), Partidos et al. (2012), Plante et al. (2011), Plante et al. (2015), Roy et al. (2014), Kim et al. (2011)
Chimeric Virus	VSVdelG-CHIKV	Glycoprotein-deficient VSV encoding CHIKV structural proteins	WT Mice	Preclinical	Chattopadhyay et al. (2012)
Chimeric Virus	VEE/CHIKV; VEE/IRES-C/CHIKV	Chimera consisting of VEEV nsP and CHIKV structural polyproteins	WT, IFN-deficient mice	Preclinical	Wang et al. (2011a, 2008)
Chimeric Virus	EILV/CHIKV	Viral genome encodes Eilat virus nonstructural and CHIKV structural proteins	Mouse, NHP	Preclinical	Erasmus et al. (2015, 2016)
Chimeric Virus	CAdVax-CHIK	Replication-deficient Ad5 encoding CHIKV structural polyprotein	WT Mice	Preclinical	Wang et al. (2011b)

(continued)

Table 1 (continued)

Platform	Name	Description	Models	Development	References
Chimeric Virus	MVA-6KE1E3E2	MVA encoding CHIKV structural polyprotein	IFN-deficient Mice	Preclinical	van den Doel et al. (2014)
Chimeric Virus	MVA-E2E3	MVA encoding CHIKV E2E3	IFN-deficient Mice	Preclinical	Weger-Lucarelli et al. (2014)
Chimeric Virus	MVA-CHIKV	MVA encoding CHIKV structural polyprotein	WT Mice	Preclinical	García-Arriaza et al. (2014)
Chimeric Virus	MV/Schwarz	Measles virus encoding CHIKV structural proteins	Mouse, NHP, Human	Phase II	Reisinger et al. (2019, Brandler et al. (2013), Ramsauer et al. (2015), Rossi et al. (2019)
Chimeric Virus	CHIK001	Chimpanzee adenovirus expressing CHIKV structural proteins	Mouse	Phase I	Ewer et al. (2017)
Inactivated Particle	UV-CHIKV	CHIKV particle inactivated by UV-irradiation	NHP	Preclinical	Nakao and Hotta (1973)
Inactivated Particle	BEI-CHIKV	CHIKV particle inactivated by binary ethyleneimine	WT Mouse	Preclinical	Gardner et al. (2010)
Inactivated Particle	Formalin-inactivated CHIKV	CHIKV particle inactivated by formalin	Mouse, Human	Preclinical	Tiwari et al. (2009)
Inactivated Particle	Tween/Ether-CHIKV	CHIKV particle inactivated by Tween 80 and ether	WT mice	Preclinical	Eckels et al. (1970)
Subunit	rCHIKVE1/E2	E. Coli synthesized recombinant E1 and E2 proteins	WT mice	Preclinical	Kumar et al. (2012)
Subunit	rE2p	E. Coli synthesized recombinant E2 proteins	WT mice	Preclinical	Khan et al. (2012)

(continued)

Table 1 (continued)

Platform	Name	Description	Models	Development	References
VLP	VRC-CHKVLP059-00-VP	Virus-like particle made on human HEK293 cells	Mouse, NHP, Human	Phase II	Kam et al. (2012a, b), Sun et al. (2013), Chang et al. (2014), Goo et al. (2016)
VLP	Sf21-VLP/PXVX0317	Virus-like particle made using baculovirus and Sf21 insect cells	WT mice	Phase II	Metz et al. (2013a)
DNA	p181/25-7	DNA vector expressing the full genomic sequence of attenuated 181/25 CHIKV strain	WT mice	Preclinical	Tretyakova et al. (2014)
DNA	CVM1-IgG	DNA vector expressing E2-reactive monoclonal antibody	WT mice	Preclinical	Muthumani et al. (2016)
DNA	pVax1-MCE321	DNA vector expressing CHIKV E1, E2, E3 proteins	Mouse, NHP	Preclinical	Muthumani et al. (2008, 2016), Mallilankaraman et al. (2011)
DNA	DREP-Env	DNA replicon encoding nsP and E1, E2, E3	WT mice	Preclinical	Hallengärd et al. (2014a, b)
mRNA	mRNA-1944	Liposome-encapsulated mRNA encoding CHIKV-directed monoclonal antibody	WT, IFN-deficient mice, NHP	Preclinical	Kose et al. (2019)
mRNA	mRNA-1388	Liposome-encapsulated mRNA encoding CHIKV structural antigens	Mouse, Human	Phase I	Shaw et al. (2019)

clinical trials have also been completed or are currently underway for some technologies. However, despite this, regulatory bodies are yet to approve the widespread commercial utilization of any vaccine against CHIKV in humans. Unfortunately, the financial cost of developing a vaccine ready for distribution in human populations is on the order of hundreds of millions of dollars and thus requires a substantial commitment from private, nonprofit, or public institutions. As such, it will be necessary that threshold levels of either commercial potential or public need are crossed and associated with suitable risk criteria in order to attract the resources necessary for bringing a CHIKV vaccine into clinical use. Given that the burden of the virus is disproportionately incurred by economically developing nations, market forces may not forecast a return on investment necessary to incentivize for-profit entities in this endeavor. Moreover, the focal and episodic nature of CHIKV outbreaks, as well as establishment of herd immunity, renders phase III efficacy trials logistically problematic. However, spontaneous and unforeseen factors that drive arbovirus emergence and spread may fluctuate in a manner that results in outbreaks of magnitude or that shift geographically such that expanded market potential is revealed. Additionally, a market may also exist among travelers from developed countries interested in obtaining immune protection prior to the visitation of afflicted areas. Military needs may also drive vaccine development and thus government(s) may fund clinical advancement in the absence of commercial demand. The Global Alliance for Vaccines and Immunization (GAVI) also financially supports efforts in low-to middle-income countries to develop vaccines based primarily on disease burden and overall program sustainability.

Priority populations in which enhanced susceptibility to or severity of CHIKV-induced disease are evident include neonates (Charlier et al. 2017) and the aged (Lang et al. 2017; Godaert et al. 2017). As such, vaccine platforms pursued clinical use need to account for the immature or senescent immune status of these groups. Specifically, this includes an enhanced safety profile (including reactogenicity, adverse effects, and vaccine-induced disease) and appropriate immunostimulatory capacity. As discussed above, LAV are considered less safe due to the possibility of reversion or enhanced replication in the absence of robust immunity. Subunit and VLP-based vaccines exhibit improved safety profiles but are clearly less immunogenic. However, adjuvant classes and combinations exist and are being actively pursued that are capable of addressing this (Bonam et al. 2017; Aspinall et al. 2007; Levy et al. 2013; Borriello et al. 2017). Importantly, clinical trials examining vaccine safety and efficacy should be designed and executed in accordance with these goals. In addition, the establishment of a standard antibody neutralization protocol and criteria that can be applied to allow comparison across studies will also be crucial for further development. Finally, prioritization of a CHIKV vaccine will also depend on the emergence and outbreaks of other viral diseases such as Zika and Ebola or unknown zoonotic agents. These impose their own unique burdens on human populations and independently attract attention, effort, and investment.

References

Akahata W, Yang Z-Y, Andersen H, Sun S, Holdaway HA, Kong W-P, Lewis MG, Higgs S, Rossmann MG, Rao S, Nabel GJ (2010) A virus-like particle vaccine for epidemic Chikungunya virus protects nonhuman primates against infection. Nature Medicine 16:334–338

Aspinall R, Del Giudice G, Effros RB, Grubeck-Loebenstein B, Sambhara S (2007) Challenges for vaccination in the elderly. Immun Ageing 4:9

Auerswald H, Boussioux C, In S, Mao S, Ong S, Huy R, Leang R, Chan M, Duong V, Ly S, Tarantola A, Dussart P (2018) Broad and long-lasting immune protection against various Chikungunya genotypes demonstrated by participants in a cross-sectional study in a Cambodian rural community. Emerg Microbes Infect 7:13

Bloch D (2017) The Cost and burden of Chikungunya in the Americas 1–34

Bonam SR, Partidos CD, Halmuthur SKM, Muller S (2017) An overview of novel adjuvants designed for improving vaccine efficacy. Trends Pharmacol Sci 38:771–793

Borriello F, Pietrasanta C, Lai JCY, Walsh LM, Sharma P, O'Driscoll DN, Ramirez J, Brightman S, Pugni L, Mosca F, Burkhart DJ, Dowling DJ, Levy O (2017) Identification and characterization of stimulator of interferon genes as a robust adjuvant target for early life immunization. Front Immunol 8:1772

Brandler S, Ruffié C, Combredet C, Brault J-B, Najburg V, Prevost M-C, Habel A, Tauber E, Desprès P, Tangy F (2013) A recombinant measles vaccine expressing Chikungunya virus-like particles is strongly immunogenic and protects mice from lethal challenge with Chikungunya virus. Vaccine 31:3718–3725

Broeckel R, Fox JM, Haese N, Kreklywich CN, Sukulpovi-Petty S, Legasse A, Smith PP, Denton M, Corvey C, Krishnan S, Colgin LMA, Ducore RM, Lewis AD, Axthelm MK, Mandron M, Cortez P, Rothblatt J, Rao E, Focken I, Carter K, Sapparapau G, Crowe JE, Diamond MS, Streblow DN (2017) Therapeutic administration of a recombinant human monoclonal antibody reduces the severity of Chikungunya virus disease in rhesus macaques. PLoS Negl Trop Dis 11:e0005637

Cardiel MH, Pons-Estel BA, Sacnun MP, Wojdyla D, Saurit V, Marcos JC, Pinto MRC, Cordeiro de Azevedo AB, da Silveira IG, Radominski SC, Ximenes AC, Massardo L, Ballesteros F, Rojas-Villarraga A, Oñate RV, Hernandez MP, Esquivel-Valerio JA, García-De La Torre I, Khoury VJ, Millán A, Soriano ER, GLADAR (2012) Treatment of early rheumatoid arthritis in a multinational inception cohort of Latin American patients: the GLADAR experience. J Clin Rheumatol 18:327–335

Chang L-J, Dowd KA, Mendoza FH, Sounders JG, Sitar S, Plummer SH, Yamshchikov G, Sarwar UN, Hu Z, Enama ME, Bailer RT, Koup RA, Schwartz RM, Akahata W, Nabel GJ, Mascola JR, Pierson TC, Graham BS, Ledgerwood JE, Team TV3S (2014) Safety and tolerability of Chikungunya virus-like particle vaccine in healthy adults: a phase 1 dose-escalation trial. The Lancet 384:2046–2052

Charlier C, Beaudoin M-C, Couderc T, Lortholary O, Lecuit M (2017) Arboviruses and pregnancy: maternal, fetal, and neonatal effects. Lancet Child Adolesc Health 1:134–146

Chattopadhyay A, Wang E, Seymour R, Weaver SC, Rose JK (2012) A chimeric vesiculo/alphavirus is an effective Alphavirus Vaccine. J Virol

Chen C-I, Clark DC, Pesavento P, Lerche NW, Luciw PA, Reisen WK, Brault AC (2010) Comparative pathogenesis of epidemic and enzootic Chikungunya viruses in a pregnant Rhesus macaque model. Am J Trop Med Hyg 83:1249–1258

Chu H, Das SC, Fuchs JF, Suresh M, Weaver SC, Stinchcomb DT, Partidos CD, Osorio JE (2013) Deciphering the protective role of adaptive immunity to CHIKV/IRES a novel candidate vaccine against Chikungunya in the A129 mouse model. Vaccine 31:3353–3360

Chua C-L, Sam I-C, Merits A, Chan Y-F (2016) Antigenic variation of East/Central/South African and Asian Chikungunya virus genotypes in neutralization by immune Sera. PLoS Negl Trop Dis 10:e0004960

Coffman RL, Sher A, Seder RA (2010) Vaccine adjuvants: putting innate immunity to work. Immunity 33:492–503

Coller B-AG, Blue J, Das R, Dubey S, Finelli L, Gupta S, Helmond F, Grant-Klein RJ, Liu K, Simon J, Troth S, VanRheenen S, Waterbury J, Wivel A, Wolf J, Heppner DG, Kemp T, Nichols R, Monath TP (2017) Clinical development of a recombinant Ebola vaccine in the midst of an unprecedented epidemic. Vaccine 35:4465–4469

Combredet C, Labrousse V, Mollet L, Lorin C, Delebecque F, Hurtrel B, McClure H, Feinberg MB, Brahic M, Tangy F (2003) A molecularly cloned Schwarz strain of measles virus vaccine induces strong immune responses in macaques and transgenic mice. J Virol 77:11546–11554

Couderc T, Chretien F, Schilte C, Disson O, Brigitte M (2008a) A mouse model for Chikungunya: young age and inefficient type-I interferon signaling are risk factors for severe disease. PLoS …

Couderc T, Chrétien F, Schilte C, Disson O, Brigitte M, Guivel-Benhassine F, Touret Y, Barau G, Cayet N, Schuffenecker I, Desprès P, Arenzana-Seisdedos F, Michault A, Albert ML, Lecuit M (2008b) A mouse model for Chikungunya: young age and inefficient type-I interferon signaling are risk factors for severe disease. PLoS Pathog 4:e29

Couderc T, Khandoudi N, Grandadam M, Visse C, Gangneux N, Bagot S, Prost J-F, Lecuit M (2009) Prophylaxis and therapy for Chikungunya virus infection. J Infect Dis 200:516–523

de Swart RL, DeSwart RL (2009) Measles studies in the macaque model. Curr Top Microbiol Immunol 330:55–72

de Brito CAA, Sohsten AKAV, Leitão CC de S, Brito R de CCM de, Valadares LDDA, Fonte CAMD, Mesquita ZB de, Cunha RV, Luz K, Leão HMC, Brito CM de, Frutuoso LCV (2016) Pharmacologic management of pain in patients with Chikungunya: a guideline. Rev Soc Bras Med Trop 49:668–679

Diallo D, Sall AA, Buenemann M, Chen R, Faye O, Diagne CT, Faye O, Ba Y, Dia I, Watts D, Weaver SC, Hanley KA, Diallo M (2012) Landscape ecology of sylvatic chikungunya virus and mosquito vectors in southeastern Senegal. PLoS Negl Trop Dis 6:e1649

Draper SJ, Cottingham MG, Gilbert SC (2013) Utilizing poxviral vectored vaccines for antibody induction-progress and prospects. Vaccine 31:4223–4230

Eckels KH, Harrison VR, Hetrick FM (1970) Chikungunya virus vaccine prepared by Tween-ether extraction. Appl Microbiol 19:321–325

Edelman R, Tacket CO, Wasserman SS, Bodison SA, Perry JG, Mangiafico JA (2000) Phase II safety and immunogenicity study of live Chikungunya virus vaccine TSI-GSD-218. Am J Trop Med Hyg 62:681–685

Enserink M (2007) Infectious diseases. Chikungunya: no longer a third world disease. Science. American Association for the Advancement of Science

Erasmus JH, Needham J, Raychaudhuri S, Diamond MS, Beasley DWC, Morkowski S, Salje H, Fernandez-Salas I, Kim DY, Frolov I, Nasar F, Weaver SC (2015) Utilization of an eilat virus-based chimera for serological detection of Chikungunya infection. PLoS Negl Trop Dis 9:e0004119

Erasmus JH, Auguste AJ, Kaelber JT, Luo H, Rossi SL, Fenton K, Leal G, Kim DY, Chiu W, Wang T, Frolov I, Nasar F, Weaver SC (2016) A Chikungunya fever vaccine utilizing an insect-specific virus platform. Nat Med 23:192–199

Ewer K, Sebastian S, Spencer AJ, Gilbert S, Hill AVS, Lambe T (2017) Chimpanzee adenoviral vectors as vaccines for outbreak pathogens. Hum Vaccin Immunother 13:3020–3032

Fox JM, Long F, Edeling MA, Lin H, van Duijl-Richter MKS, Fong RH, Kahle KM, Smit JM, Jin J, Simmons G, Doranz BJ, Crowe JE Jr, Fremont DH, Rossmann MG, Diamond MS (2015) Broadly neutralizing alphavirus antibodies bind an epitope on E2 and inhibit entry and egress. Cell 1–14

Galatas B, Ly S, Duong V, Baisley K, Nguon K, Chan S, Huy R, Ly S, Sorn S, Som L, Buchy P, Tarantola A (2016) Long-lasting immune protection and other epidemiological findings after Chikungunya emergence in a Cambodian rural community, April 2012. PLoS Negl Trop Dis 10:e0004281

García-Arriaza J, Cepeda V, Hallengärd D, Sorzano CÓS, Kümmerer BM, Liljeström P, Esteban M (2014) A novel poxvirus-based vaccine, MVA-CHIKV, is highly immunogenic and protects mice against Chikungunya infection. J Virol 88:3527–3547

Gardner J, Anraku I, Le TT, Larcher T, Major L, Roques P, Schroder WA, Higgs S, Suhrbier A (2010) Chikungunya virus arthritis in adult wild-type mice. J Virol 84:8021–8032

Gardner CL, Burke CW, Higgs ST, Klimstra WB, Ryman KD (2012) Interferon-alpha/beta deficiency greatly exacerbates arthritogenic disease in mice infected with wild-type chikungunya virus but not with the cell culture-adapted live-attenuated 181/25 vaccine candidate. Virology 425:103–112

Gardner CL, Hritz J, Sun C, Vanlandingham DL, Song TY, Ghedin E, Higgs S, Klimstra WB, Ryman KD (2014) Deliberate attenuation of chikungunya virus by adaptation to heparan sulfate-dependent infectivity: a model for rational arboviral vaccine design. PLoS Negl Trop Dis 8:e2719

Genzel Y (2015) Designing cell lines for viral vaccine production: Where do we stand? Biotechnol J 10:728–740

Godaert L, Najioullah F, Bartholet S, Colas S, Yactayo S, Cabié A, Fanon J-L, Césaire R, Dramé M (2017) Atypical clinical presentations of acute phase Chikungunya virus infection in older adults. J Am Geriatr Soc 65:2510–2515

Goo L, Dowd KA, Lin T-Y, Mascola JR, Graham BS, Ledgerwood JE, Pierson TC (2016) A virus-like particle vaccine elicits broad neutralizing antibody responses in humans to All Chikungunya virus genotypes. J Infect Dis 214:1487–1491

Gorchakov R, Wang E, Leal G, Forrester NL, Plante K, Rossi SL, Partidos CD, Adams AP, Seymour RL, Weger J, Borland EM, Sherman MB, Powers AM, Osorio JE, Weaver SC (2012) Attenuation of Chikungunya virus vaccine strain 181/clone 25 is determined by two amino acid substitutions in the E2 envelope glycoprotein. J Virol 86:6084–6096

Haese NN, Broeckel RM, Hawman DW, Heise MT, Morrison TE, Streblow DN (2016) Animal models of Chikungunya virus infection and disease. J Infect Dis 214:S482–S487

Haglund K, Leiner I, Kerksiek K, Buonocore L, Pamer E, Rose JK (2002) High-level primary CD8 (+) T-cell response to human immunodeficiency virus type 1 gag and env generated by vaccination with recombinant vesicular stomatitis viruses. J Virol 76:2730–2738

Hallengard D, Lum FM, Kummerer BM, Lulla A, Lulla V, Garcia-Arriaza J, Fazakerley JK, Roques P, Le Grand R, Merits A, Ng LFP, Esteban M, Liljestrom P (2014) Prime-boost immunization strategies against Chikungunya virus. J Virol

Hallengärd D, Kakoulidou M, Lulla A, Kümmerer BM, Johansson DX, Mutso M, Lulla V, Fazakerley JK, Roques P, Le Grand R, Merits A, Liljeström P (2014) Novel attenuated Chikungunya vaccine candidates elicit protective immunity in C57BL/6 mice. J Virol 88:2858–2866

Harrison VR, Eckels KH, Bartelloni PJ, Hampton C (1971) Production and evaluation of a formalin-killed Chikungunya vaccine. J Immunol 107:643–647

Hawman DW, Stoermer KA, Montgomery SA, Pal P, Oko L, Diamond MS, Morrison TE (2013) Chronic joint disease caused by persistent Chikungunya virus infection is controlled by the adaptive immune response. J Virol 87:13878–13888

Hoarau JJ, Gay F, Pellé O, Samri A, Jaffar-Bandjee MC, Gasque P, Autran B (2013) Identical strength of the T cell responses against E2, nsP1 and capsid CHIKV proteins in recovered and chronic patients after the epidemics of 2005–2006 in La Reunion Island. PLoS ONE 8:e84695

Hu M-M, Shu H-B (2018) Cytoplasmic mechanisms of recognition and defense of microbial nucleic acids. Annu Rev Cell Dev Biol

Iwasaki A, Medzhitov R (2015) Control of adaptive immunity by the innate immune system. Nat Immunol 16:343–353

Kam Y-W, Lee WWL, Simarmata D, Harjanto S, Teng T-S, Tolou H, Chow A, Lin RTP, Leo Y-S, REnia L, Ng LFP (2012) Longitudinal analysis of the human antibody response to Chikungunya virus infection: implications for serodiagnosis and vaccine development. J Virol 86:13005–13015

Kam Y-W, Simarmata D, Chow A, Her Z, Teng T-S, Ong EKS, REnia L, Leo Y-S, Ng LFP (2012a) Early appearance of neutralizing immunoglobulin G3 antibodies is associated with Chikungunya virus clearance and long-term clinical protection. J Infect Dis 205:1147–1154

Kam Y-W, Lum F-M, Teo T-H, Lee WWL, Simarmata D, Harjanto S, Chua C-L, Chan Y-F, Wee J-K, Chow A, Lin RTP, Leo Y-S, Le Grand R, Sam I-C, Tong J-C, Roques P, Wiesmüller K-H, REnia L, Rötzschke O, Ng LFP (2012b) Early neutralizing IgG response to Chikungunya virus in infected patients targets a dominant linear epitope on the E2 glycoprotein. EMBO Mol Med 4:330–343

Khan M, Dhanwani R, Rao PVL, Parida M (2012) Subunit vaccine formulations based on recombinant envelope proteins of Chikungunya virus elicit balanced Th1/Th2 response and virus-neutralizing antibodies in mice. Virus Research 167:236–246

Kim DY, Atasheva S, Foy NJ, Wang E, Frolova EI, Weaver S, Frolov I (2011) Design of chimeric alphaviruses with a programmed, attenuated, cell type-restricted phenotype. J Virol 85:4363–4376

Kose N, Fox JM, Sapparapu G, Bombardi R, Tennekoon RN, de Silva AD, Elbashir SM, Theisen MA, Humphris-Narayanan E, Ciaramella G, Himansu S, Diamond MS, Crowe JE (2019) A lipid-encapsulated mRNA encoding a potently neutralizing human monoclonal antibody protects against Chikungunya infection. Sci Immunol 4:eaaw6647

Kumar M, Sudeep AB, Arankalle VA (2012) Evaluation of recombinant E2 protein-based and whole-virus inactivated candidate vaccines against Chikungunya virus. Vaccine 30:6142–6149

Lanciotti RS, Valadere AM (2014) Transcontinental movement of Asian genotype Chikungunya virus. Emerging Infect Dis 20:1400–1402

Lang PO, Loulergue P, Aspinall R (2017) Chikungunya Virus Infection: Why Should U.S. Geriatricians Be Aware of It? J Am Geriatr Soc 65:2529–2534

Langsjoen RM, Haller SL, Roy CJ, Vinet-Oliphant H, Bergren NA, Erasmus JH, Livengood JA, Powell TD, Weaver SC, Rossi SL (2018) Chikungunya virus strains show lineage-specific variations in virulence and cross-protective ability in murine and nonhuman primate models. mBio 9:e02449–17–13

Leparc-Goffart I, Nougairede A, Cassadou S, Prat C, de Lamballerie X (2014) Chikungunya in the Americas. Lancet 383:514

Levitt NH, Ramsburg HH, Hasty SE, Repik PM, Cole FE, Lupton HW (1986) Development of an attenuated strain of Chikungunya virus for use in vaccine production. Vaccine 4:157–162

Levy O, Goriely S, Kollmann TR (2013) Immune response to vaccine adjuvants during the first year of life. Vaccine 31:2500–2505

Liu MA (2010) Immunologic basis of vaccine vectors. Immunity 33:504–515

Lloyd J, Cheyne J (2017) The origins of the vaccine cold chain and a glimpse of the future. Vaccine 35:2115–2120

Lum F-M, Teo T-H, Lee WWL, Kam Y-W, REnia L, Ng LFP (2013) An essential role of antibodies in the control of Chikungunya virus infection. J Immunol 190:6295–6302

Majhen D, Calderon H, Chandra N, Fajardo CA, Rajan A, Alemany R, Custers J (2014) Adenovirus-based vaccines for fighting infectious diseases and cancer: progress in the field. Human Gene Ther 25:301–317

Mallilankaraman K, Shedlock DJ, Bao H, Kawalekar OU, Fagone P, Ramanathan AA, Ferraro B, Stabenow J, Vijayachari P, Sundaram SG, Muruganandam N, Sarangan G, Srikanth P, Khan AS, Lewis MG, Kim JJ, Sardesai NY, Muthumani K, Weiner DB (2011) A DNA vaccine against Chikungunya virus is protective in mice and induces neutralizing antibodies in mice and nonhuman primates. PLoS Negl Trop Dis 5:e928

Messaoudi I, Vomaske J, Totonchy T, Kreklywich CN, Haberthur K, Springgay L, Brien JD, Diamond MS, Defilippis VR, Streblow DN (2013) Chikungunya virus infection results in higher and persistent viral replication in aged rhesus macaques due to defects in anti-viral immunity. PLoS Negl Trop Dis 7:e2343

Metz SW, Gardner J, Geertsema C, Le TT, Goh L, Vlak JM, Suhrbier A, Pijlman GP (2013a) Effective Chikungunya virus-like particle vaccine produced in insect cells. PLoS Negl Trop Dis 7:e2124

Metz SW, Martina BE, van den Doel P, Geertsema C, Osterhaus AD, Vlak JM, Pijlman GP (2013b) Chikungunya virus-like particles are more immunogenic in a lethal AG129 mouse model compared to glycoprotein E1 or E2 subunits. Vaccine 31:6092–6096

Mohsen MO, Zha L, Cabral-Miranda G, Bachmann MF (2017) Major findings and recent advances in virus-like particle (VLP)-based vaccines. Semin Immunol 34:123–132

Morrison TE, Oko L, Montgomery SA, Whitmore AC, Lotstein AR, Gunn BM, Elmore SA, Heise MT (2011) A mouse model of Chikungunya virus-induced musculoskeletal inflammatory disease. AJPA 178:32–40

Moss B, Carroll MW, Wyatt LS, Bennink JR, Hirsch VM, Goldstein S, Elkins WR, Fuerst TR, Lifson JD, Piatak M, Restifo NP, Overwijk W, Chamberlain R, Rosenberg SA, Sutter G (1996) Host range restricted, non-replicating vaccinia virus vectors as vaccine candidates. Adv Exp Med Biol 397:7–13

Mühlebach MD (2017) Vaccine platform recombinant measles virus. Virus Genes 53:733–740

Muthumani K, Lankaraman KM, Laddy DJ, Sundaram SG, Chung CW, Sako E, Wu L, Khan A, Sardesai N, Kim JJ, Vijayachari P, Weiner DB (2008) Immunogenicity of novel consensus-based DNA vaccines against Chikungunya virus. Vaccine 26:5128–5134

Muthumani K, Block P, Flingai S, Muruganantham N, Chaaithanya IK, Tingey C, Wise M, Reuschel EL, Chung C, Muthumani A, Sarangan G, Srikanth P, Khan AS, Vijayachari P, Sardesai NY, Kim JJ, Ugen KE, Weiner DB (2016) Rapid and long-term immunity elicited by DNA-encoded antibody prophylaxis and DNA vaccination against Chikungunya virus. J Infect Dis 214:369–378

Nakao E, Hotta S (1973) Immunogenicity of purified, inactivated Chikungunya virus in monkeys. Bull World Health Organ 48:559–562

Nasar F, Gorchakov RV, Tesh RB, Weaver SC (2015) Eilat virus host range restriction is present at multiple levels of the virus life cycle. J Virol (5 ed.) 89:1404–1418

Pal P, Dowd KA, Brien JD, Edeling MA, Gorlatov S, Johnson S, Lee I, Akahata W, Nabel GJ, Richter MKS, Smit JM, Fremont DH, Pierson TC, Heise MT, Diamond MS (2013) Development of a highly protective combination monoclonal antibody therapy against Chikungunya virus. PLoS Pathog 9:e1003312

Partidos CD, Paykel J, Weger J, Borland EM, Powers AM, Seymour R, Weaver SC, Stinchcomb DT, Osorio JE (2012) Cross-protective immunity against o'nyong-nyong virus afforded by a novel recombinant Chikungunya vaccine. Vaccine 30:4638–4643

Piper A, Ribeiro M, Smith KM, Briggs CM, Huitt E, Nanda K, Spears CJ, Quiles M, Cullen J, Thomas ME, Brown DT, Hernandez R (2013) Chikungunya virus host range E2 transmembrane deletion mutants induce protective immunity against challenge in C57BL/6 J Mice. J Virol 87:6748–6757

Plante K, Wang E, Partidos CD, Weger J, Gorchakov R, Tsetsarkin K, Borland EM, Powers AM, Seymour R, Stinchcomb DT, Osorio JE, Frolov I, Weaver SC (2011) Novel Chikungunya vaccine candidate with an IRES-based attenuation and host range alteration mechanism. PLoS Pathog 7:e1002142

Plante KS, Rossi SL, Bergren NA, Seymour RL, Weaver SC (2015) Extended preclinical safety, efficacy and stability testing of a live-attenuated Chikungunya Vaccine candidate. PLoS Negl Trop Dis 9:e0004007–19

Poo YS, Rudd PA, Gardner J, Wilson JAC, Larcher T, Colle M-A, Le TT, Nakaya HI, Warrilow D, Allcock R, Bielefeldt-Ohmann H, Schroder WA, Khromykh AA, Lopez JA, Suhrbier A (2014) Multiple immune factors are involved in controlling acute and chronic Chikungunya virus infection. PLoS Negl Trop Dis 8:e3354–15

Ramsauer K, Tangy F (2016) Chikungunya virus vaccines: viral vector-based approaches. J Infect Dis 214:S500–S505

Ramsauer K, Schwameis M, Firbas C, Müllner M, Putnak RJ, Thomas SJ, Desprès P, Tauber E, Jilma B, Tangy F (2015) Immunogenicity, safety, and tolerability of a recombinant measles-virus-based Chikungunya vaccine: a randomised, double-blind, placebo-controlled, active-comparator, first-in-man trial. Lancet Infect Dis 15:519–527

Reed SG, Orr MT, Fox CB (2013) Key roles of adjuvants in modern vaccines. Nat Med 19:1597–1608

Reisinger EC, Tschismarov R, Beubler E, Wiedermann U, Firbas C, Loebermann M, Pfeiffer A, Muellner M, Tauber E, Ramsauer K (2019) Immunogenicity, safety, and tolerability of the measles-vectored chikungunya virus vaccine MV-CHIK: a double-blind, randomised, placebo-controlled and active-controlled phase 2 trial. Lancet 392:2718–2727

Roberts A, Buonocore L, Price R, Forman J, Rose JK (1999) Attenuated vesicular stomatitis viruses as vaccine vectors. J Virol 73:3723–3732

Roques P, Ljungberg K, Kümmerer BM, Gosse L, Dereuddre-Bosquet N, Tchitchek N, Hallengärd D, García-Arriaza J, Meinke A, Esteban M, Merits A, Le Grand R, Liljeström P (2017) Attenuated and vectored vaccines protect nonhuman primates against Chikungunya virus. JCI Insight 2:e83527

Rossi SL, Comer JE, Wang E, Azar SR, Lawrence WS, Plante JA, Ramsauer K, Schrauf S, Weaver SC (2019) Immunogenicity and efficacy of a measles virus-vectored Chikungunya vaccine in nonhuman primates. J Infect Dis

Roy CJ, Adams AP, Wang E, Plante K, Gorchakov R, Seymour RL, Vinet-Oliphant H, Weaver SC (2014) Chikungunya vaccine candidate is highly attenuated and protects nonhuman primates against telemetrically monitored disease following a single dose. J Infect Dis 209:1891–1899

Rudd PA, Wilson J, Gardner J, Larcher T, Babarit C, Le TT, Anraku I, Kumagai Y, Loo YM, Gale M, Akira S, Khromykh AA, Suhrbier A (2012) Interferon response factors 3 and 7 protect against Chikungunya virus hemorrhagic fever and shock. J Virol 86:9888–9898

Sahadeo NSD, Allicock OM, De Salazar PM, Auguste AJ, Widen S, Olowokure B, Gutierrez C, Valadere AM, Polson-Edwards K, Weaver SC, Carrington CVF (2017) Understanding the evolution and spread of Chikungunya virus in the Americas using complete genome sequences. Virus Evol 3:vex010–10

Saxena M, Van TTH, Baird FJ, Coloe PJ, Smooker PM (2013) Pre-existing immunity against vaccine vectors–friend or foe? Microbiology (Reading, England) 159:1–11

Seyler T, Hutin Y, Ramanchandran V, Ramakrishnan R, Manickam P, Murhekar M (2010) Estimating the burden of disease and the economic cost attributable to chikungunya, Andhra Pradesh, India, 2005–2006. Trans R Soc Trop Med Hyg 104:133–138

Shaw C, Panther L, August A, Zaks T, Smolenov I, Bart S, Watson M (2019) Safety and immunogenicity of a mRNA-based Chikungunya vaccine in a phase 1 dose-ranging trial. Int J Infect Dis 79:17

Sheela PJ, Sumathy K (2013) Serological correlates of immune protection conferred by Chikungunya virus infection. Acta Virol 57:471–473

Simmons G, Brès V, Lu K, Liss NM, Brambilla DJ, Ryff KR, Bruhn R, Velez E, Ocampo D, Linnen JM, Latoni G, Petersen LR, Williamson PC, Busch MP (2016) High incidence of Chikungunya virus and frequency of viremic blood donations during epidemic, Puerto Rico, USA, 2014. Emerg Infect Dis 22:1221–1228

Smith SA, Silva LA, Fox JM, Flyak AI, Kose N, Sapparapu G, Khomandiak S, Khomadiak S, Ashbrook AW, Kahle KM, Fong RH, Swayne S, Doranz BJ, McGee CE, Heise MT, Pal P, Brien JD, Austin SK, Diamond MS, Dermody TS, Crowe JE (2015) Isolation and characterization of broad and ultrapotent human monoclonal antibodies with therapeutic activity against Chikungunya virus. Cell Host Microbe 18:86–95

Strauss JH, Strauss EG (1994) The alphaviruses: gene expression, replication, and evolution. Microbiol Rev 58:491–562

Sun S, Xiang Y, Akahata W, Holdaway H, Pal P, Zhang X, Diamond MS, Nabel GJ, Rossmann MG (2013) Structural analyses at pseudo atomic resolution of Chikungunya virus and antibodies show mechanisms of neutralization. eLife 2:e00435

Tangy F, Naim HY (2005) Live attenuated measles vaccine as a potential multivalent pediatric vaccination vector. Viral Immunol 18:317–326

Taylor A, Herrero LJ, Rudd PA, Mahalingam S (2015) Mouse models of alphavirus-induced inflammatory disease. J Gen Virol 96:221–238

Teo TH, Lum FM, Claser C, Lulla V, Lulla A, Merits A, Renia L, Ng LFP (2012a) A pathogenic role for $CD4^+$ T cells during Chikungunya virus infection in mice. J Immunol 190:259–269

Teo T-H, Lum F-M, Lee WWL, Ng LFP (2012b) Mouse models for Chikungunya virus: deciphering immune mechanisms responsible for disease and pathology. Immunol Res 53:136–147

Teo T-H, Her Z, Tan JJL, Lum F-M, Lee WWL, Chan Y-H, Ong R-Y, Kam Y-W, Leparc-Goffart I, Gallian P, REnia L, de Lamballerie X, Ng LFP (2015) Caribbean and La Réunion Chikungunya virus isolates differ in their capacity to induce pro-inflammatory Th1 and NK cell responses and acute joint pathology. J Virol JVI.00909–15–59

Thompson EA, Loré K (2017) Non-human primates as a model for understanding the mechanism of action of toll-like receptor-based vaccine adjuvants. Curr Opin Immunol 47:1–7

Tiwari M, Parida M, Santhosh SR, Khan M, Dash PK, Rao PVL (2009) Assessment of immunogenic potential of Vero adapted formalin inactivated vaccine derived from novel ECSA genotype of Chikungunya virus. Vaccine 27:2513–2522

Tretyakova I, Hearn J, Wang E, Weaver S, Pushko P (2014) DNA vaccine initiates replication of live attenuated Chikungunya virus in vitro and elicits protective immune response in mice. J Infect Dis 209:1882–1890

Tsetsarkin KA, Weaver SC (2011) Sequential adaptive mutations enhance efficient vector switching by Chikungunya virus and its epidemic emergence. PLoS Pathog 7:e1002412

Tsetsarkin K, Higgs S, McGee CE, de Lamballerie X, Charrel RN, Vanlandingham DL (2006) Infectious clones of Chikungunya virus (La Réunion isolate) for vector competence studies. Vector-Borne and Zoonotic Dis 6:325–337

Tsetsarkin KA, Vanlandingham DL, McGee CE, Higgs S (2007) A single mutation in Chikungunya virus affects vector specificity and epidemic potential. PLoS Pathog 3:e201

Tsetsarkin KA, Chen R, Weaver SC (2016) Interspecies transmission and chikungunya virus emergence. Curr Opin Virol 16:143–150

Ulmer JB, Valley U, Rappuoli R (2006) Vaccine manufacturing: challenges and solutions. Nat Biotechnol 24:1377–1383

van den Doel P, Volz A, Roose JM, Sewbalaksing VD, Pijlman GP, van Middelkoop I, Duiverman V, van de Wetering E, Sutter G, Osterhaus ADME, Martina BEE (2014) Recombinant modified vaccinia virus Ankara expressing glycoprotein E2 of Chikungunya virus protects AG129 mice against lethal challenge. PLoS Negl Trop Dis 8:e3101

van Duijl-Richter MKS, Hoornweg TE, Rodenhuis-Zybert IA, Smit JM (2015) Early events in Chikungunya virus infection-from virus cell binding to membrane fusion. Viruses 7:3647–3674

Vaney M-C, Duquerroy S, Rey FA (2013) Alphavirus structure: activation for entry at the target cell surface. Curr Opin Virol 3:151–158

Vanlandingham DL, Tsetsarkin K, Hong C, Klingler K, McElroy KL, Lehane MJ, Higgs S (2005) Development and characterization of a double subgenomic Chikungunya virus infectious clone to express heterologous genes in *Aedes aegypti* mosquitoes. Insect Biochem Mol Biol 35:1162–1170

Vasilakos JP, Tomai MA (2013) The use of toll-like receptor 7/8 agonists as vaccine adjuvants. Expert Rev Vaccines 12:809–819

Volk SM, Chen R, Tsetsarkin KA, Adams AP, Garcia TI, Sall AA, Nasar F, Schuh AJ, Holmes EC, Higgs S, Maharaj PD, Brault AC, Weaver SC (2010) Genome-scale phylogenetic analyses of Chikungunya virus reveal independent emergences of recent epidemics and various evolutionary rates. J Virol 84:6497–6504

Volz A, Sutter G (2017) Modified vaccinia virus Ankara: history, value in basic research, and current perspectives for vaccine development. Adv Virus Res 97:187–243

Voss JE, Vaney M-C, Duquerroy S, Vonrhein C, Girard-Blanc C, Crublet E, Thompson A, Bricogne G, Rey FA (2010) Glycoprotein organization of Chikungunya virus particles revealed by X-ray crystallography. Nature 468:709–712

Wang E, Volkova E, Adams AP, Forrester N, Xiao S-Y, Frolov I, Weaver SC (2008) Chimeric alphavirus vaccine candidates for Chikungunya. Vaccine 26:5030–5039

Wang E, Kim DY, Weaver SC, Frolov I (2011a) Chimeric Chikungunya viruses are nonpathogenic in highly sensitive mouse models but efficiently induce a protective immune response. J Virol 85:9249–9252

Wang D, Suhrbier A, Penn-Nicholson A, Woraratanadharm J, Gardner J, Luo M, Le TT, Anraku I, Sakalian M, Einfeld D, Dong JY (2011b) A complex adenovirus vaccine against chikungunya virus provides complete protection against viraemia and arthritis. Vaccine 29:2803–2809

Weger-Lucarelli J, Chu H, Aliota MT, Partidos CD, Osorio JE (2014) A novel MVA vectored Chikungunya virus vaccine elicits protective immunity in mice. PLoS Negl Trop Dis 8:e2970

Weger-Lucarelli J, Aliota MT, Kamlangdee A, Osorio JE (2015) Identifying the role of E2 domains on alphavirus neutralization and protective immune responses. PLoS Negl Trop Dis 9: e0004163–17

Wei Chiam C, Fun Chan Y, Chai Ong K, Thong Wong K, Sam I-C (2015) Neurovirulence comparison of chikungunya virus isolates of the Asian and East/Central/South African genotypes from Malaysia. J Gen Virol 96:3243–3254

Yoon I-K, Alera MT, Lago CB, Tac-An IA, Villa D, Fernandez S, Thaisomboonsuk B, Klungthong C, Levy JW, Velasco JM, Roque VG, Salje H, Macareo LR, Hermann LL, Nisalak A, Srikiatkhachorn A (2015) High rate of subclinical Chikungunya virus infection and association of neutralizing antibody with protection in a prospective cohort in the Philippines. PLoS Negl Trop Dis 9:e000376

Ziegler SA, Lu L, da Rosa APAT, Xiao S-Y, Tesh RB (2008) An animal model for studying the pathogenesis of Chikungunya virus infection. Am J Trop Med Hyg 79:133–139

Small Molecule Inhibitors Targeting Chikungunya Virus

Nicole Haese, John Powers and Daniel N. Streblow

Contents

Abstract Chikungunya virus (CHIKV) infection in humans is rarely fatal but is often associated with chronic joint and muscle pain. Chronic CHIKV disease is highly debilitating and is associated with viral persistence. To date, there are no approved vaccines or therapeutics to prevent or treat CHIKV infections once they are established. Current palliative treatments aim to reduce joint inflammation and pain associated with acute and chronic CHIKV disease. Development of novel

N. Haese · J. Powers · D. N. Streblow (✉)
Vaccine and Gene, Therapy Institute and Oregon National Primate Research Center, Oregon Health & Science University, 505 NW 185th Avenue, 97006 Beaverton, OR, USA
e-mail: streblow@ohsu.edu

Current Topics in Microbiology and Immunology (2022) 435: 107–139
https://doi.org/10.1007/82_2020_195

Published online: 24 January 2019

therapeutics that reduces viral loads should positively impact virus inflammatory disease and improve patient outcomes following CHIKV infection. Therapies that target multiple aspects of CHIKV replication cycle should be developed since the virus is capable of rapidly mutating around any single therapeutic. This review summarizes the current status of small molecule inhibitor development against CHIKV.

1 Introduction

Chikungunya virus (CHIKV) is an alphavirus of the Togaviridae family transmitted to humans by *Aedes* mosquitoes. Since CHIKV was first isolated, the virus has spread causing multiple endemic and large epidemic outbreaks. Virus sequencing has identified the emergence of four different virus lineages driven, in part, by adaptation of the virus to *A. albopictus* mosquitoes during the 2000s outbreak. This adaptation allowed virus transmission into more temperate climates (Schuffenecker et al. 2006; Tsetsarkin et al. 2007). Interest in CHIKV has been driven by the recent re-emergence of the virus. CHIKV emerged on a global level starting on the coast of Kenya in 2004, in a wave that continued to spread across the islands of the Indian Oceans. This spread continued to Asia before arriving in the Caribbean region in 2013, Brazil in 2014 and then the rest of the American continents (Charrel et al. 2007; Weaver and Lecuit 2015; Yang et al. 2017). During this time, CHIKV outbreaks were also reported in Italy and France due to the emergence of the ability to use *A. albopictus* mosquitoes as a transmission vector (Cassadou et al. 2014; Rezza et al. 2007; Venturi et al. 2017). While the most recent reported outbreak of CHIKV was in Mombasa, Kenya in February 2018 (WHO Web site), this cycle of emergence and global spread is likely to repeat as antiviral immunity wanes at the population level.

Infection with CHIKV causes a febrile illness that is highly associated with severe joint/muscle pain and is often accompanied by a rash. For some individuals, the infection is self-limiting and symptoms resolve within a few weeks following infection. However, about 40–80% of those infected will continue to experience chronic joint and muscle pain that can last for months to years after CHIKV infection (Borgherini et al. 2008; Gerardin et al. 2008). Viral RNA can be detected long term in these patients, but the precise mechanisms that contribute to the development of chronic CHIKV disease are still unclear (Chopra et al. 2012; Simon et al. 2015; Sissoko et al. 2009). Atypical complications of CHIKV infection may occur in the elderly and in individuals with comorbidities, such as encephalitis (Lebrun et al. 2009). Fetal CHIKV infections during pregnancy are rare, but perinatal infection of newborns occurs during the intrapartum period, resulting in a high mortality rate and those surviving may suffer severe, lifelong neurological outcomes (Bandeira et al. 2016; Gopakumar et al. 2012; Lyra et al. 2016).

Currently, there is no licensed virus-specific treatment to prevent or treat CHIKV infection and disease. Extensive time and effort have gone into the development of multiple anti-CHIKV treatment platforms including vaccines, gene and immunotherapeutic strategies, as well as small molecule inhibitors. Herein, we focus on recent advances in the development of small molecule inhibitors directed against CHIKV replication as treatments for viral infection and disease.

2 CHIKV Genome Organization and Replication Cycle

CHIKV contains a single-stranded, positive-sense linear genome that is approximately 11.8 kb in length (Fig. 1) (Solignat et al. 2009). The genome contains two open reading frames (ORFs), and unique polyproteins are generated from two individual RNA species (Simmons and Strauss 1972; Strauss et al. 1984). The non-structural proteins (nsP1-4) are created from full-length, genomic mRNA, whereas the structural polyprotein containing capsid, E3, E2, 6K, and E1 are synthesized from a subgenomic mRNA (sgmRNA) species. Both genomic and sgmRNAs are flanked by a 5′ 7-methylguanosine cap and a 3′ polyadenylated tail (Dubin and Stollar 1975; Dubin et al. 1977). There are three untranslated regions (UTRs) within the genome; individual UTRs present at the 5′ and 3′ ends, and the third UTR is located between the ORFs, which contains the promoter sequence present on the sgmRNA that is required for translation of the structural polyprotein.

The CHIKV replication cycle is shown in Fig. 2. The Chikungunya virion contains 80 trimeric spikes of E1 and E2 glycoprotein heterodimers on the surface (Voss et al. 2010). E2 facilitates attachment to host cells followed by internalization through clathrin-mediated endocytosis (CME). Within the early endosomes, the E1 protein facilitates pH-dependent fusion of the viral envelope with the endosomal membrane (van Duijl-Richter et al. 2015). This process is followed by disassembly of the virus nucleocapsid and release of the viral genome into the cytosol. The CHIKV nsPs are synthesized by the host cell translation machinery creating the precursor polyproteins P123 and P1234. The majority of the translation products are

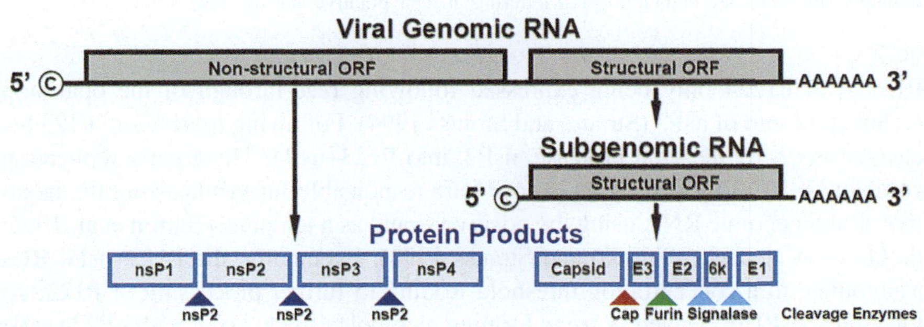

Fig. 1 Schematic representation of the Chikungunya virus genome. The CHIKV genomic and subgenomic RNAs are depicted. Following translation, the two polyproteins are cleaved by viral- or host-specific proteases to release the mature forms of the viral proteins

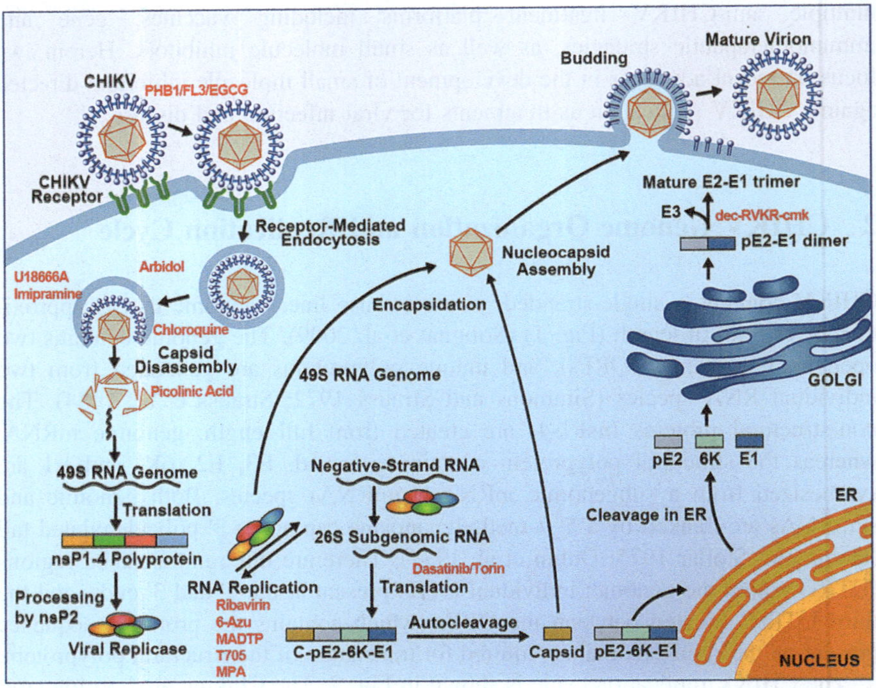

Fig. 2 Chikungunya virus replication cycle. CHIKV enters via receptor-mediated endocytosis following the binding of the envelope protein to cellular receptors (i.e., Mxra8). Low pH triggers virus particle release from the endosome. Within the cytoplasm, the virus capsid is disassembled, which allows translation of the nonstructural ORF from the viral genomic RNA. Processed nonstructural proteins form a replication complex that synthesizes both the negative strand genomic length RNA that is used as a template for the production of genomic and subgenomic RNA species. Translation of the subgenomic RNA produces the structural proteins that are processed into their mature forms required for encapsidation of the genomic RNA strand and nucleocapsid assembly. Viral envelope proteins are processed and modified in the endoplasmic reticulum and Golgi apparatus. At the plasma membrane, the glycoproteins are loaded onto viral nucleocapsids during envelopment and release at the plasma membrane. A subset of known small molecule antivirals are depicted in red lettering at their putative site of effect

P123 with P1234 only being expressed following read-through of the opal stop codon at the end of nsP3 (Strauss and Strauss 1994). Following translation, P1234 is cleaved in *cis* by the viral protease, nsP2, into P123+nsP4. These early replication complexes (RCs) formed by P123+nsP4 are responsible for synthesizing the negative strand genomic RNA using the positive strand as a template (Barton et al. 1991; de Groot et al. 1990; Shirako and Strauss 1994). Eventually, the P123+nsP4 RCs accumulate to a concentration threshold leading to further processing of P123. At this point, nsP1 is cleaved in *trans* forming a complex made up of nsP1+P23+nsP4 that for a short time, is capable of both negative and positive strand synthesis. Subgenomic viral RNA synthesis (Jose et al. 2009; van der Heijden and Bol 2002; Vasiljeva et al. 2003) occurs following cleavage of the P2/3 intermediate resulting in fully processed nsPs, forming stable late RCs capable of synthesizing both the full-length positive-sense genomic RNA and the sgmRNA (Kim et al. 2004). CHIKV replication induces the formation of host cell membrane invaginations called spherules. Spherules are connected to the cytoplasm by a narrow neck wherein early RCs are thought to localize with newly synthesized negative strand RNA. Double-stranded RNA intermediates are located in the interior of the spherule where they are protected from degradation and detection by innate immune sensors (Frolova et al. 2010; Utt et al. 2016). As the infection progresses a portion of the spherules will become internalized to form a virus-induced type 1 cytopathic vacuole (CPV) near the endoplasmic reticulum. The formation of CPVs is triggered by activation of the PI3K-AKT-mTOR pathway and is made up of membranes from both the endosome and lysosome (Spuul et al. 2010; Thaa et al. 2015).

The viral sgmRNA is translated into the structural polyprotein (capsid-E3-E2-6k-E1). Once formed the capsid protein undergoes immediate autocleavage, while the other proteins are still in the process of translation (Thomas et al. 2010). The cleaved capsid binds newly synthesized genomic viral RNA and facilitates the formation of the nucleocapsid core. The remaining structural protein polypeptide is cleaved by host proteases into pE2 (E3-E2), 6K, and E1. E1 and pE2 form non-covalent heterodimers that travel through the Golgi secretory pathway where they are post-translationally modified. Within the Golgi, the host enzyme furin cleaves pE2 into mature viral E2 and E3 (Ozden et al. 2008). The processed glycoproteins are transported to the plasma membrane where they are inserted into the host cell plasma membrane. At the same time, nucleocapsid cores containing infectious viral RNA genomes are recruited to the host cell plasma membrane for envelopment as the particles bud from the cell, thus completing the replication cycle (Garoff et al. 2004; Jose et al. 2012; Strauss and Strauss 1994).

3 Methods for Identifying Small Molecule Inhibitors Against CHIKV

The first step in many workflows for the identification of small molecule antivirals is the development and validation of a high-throughput screening (HTS) platform. A variety of CHIKV antiviral screening methods have been developed such as those based on a reduction in cellular cytopathic effect (CPE), changes in phenotype, reduction in replicon/minigenome readouts, as well as in silico virtual screening that requires solved viral protein structures.

Most conventional antiviral HTS campaigns utilize cell-based screening with readouts to measure CPE. Similar to other viruses, CHIKV infection causes robust CPE that can be easily measured as an increase in cell survival for those compounds demonstrating antiviral activity. Since some compounds in libraries can exhibit cytostatic and/or cytotoxicity, CPE-reduction-based antiviral screens provide two simultaneous readouts by identifying compounds with antiviral activity and those with low inherent cytotoxicity. Several parameters dictate the overall effectiveness and reproducibility of CPE-based HTS assays including cell type and passage number, virus strain, multiplicity of infection, and time point of development as well as the specific readout. Modifying these culture and assay conditions allows for optimal virus growth and identifiable CPE. In CHIKV CPE assays, cell types such as Vero, *A. albopictus* clone (C6/36), telomerized human fibroblasts, and Baby Hamster Kidney fibroblasts (BHK-21) cells have been successfully utilized to identify antiviral compounds (Bhat et al. 2019). Cell viability assay readouts previously used for CHIKV HTS include measuring cellular dehydrogenase activity using the colorimetric dyes MTT, MTS, or XTT; mitochondrial membrane potential using fluorescent mitochondrial tracker dyes; intracellular esterase cleavage fluorescent dyes; ATP-based luciferase reagents; and those assays that measure oxygen consumption or glycolysis. While these assays indirectly measure viral infection through cell viability, direct detection of viral protein production using viral-specific antibodies or tagged viruses allows one to directly assess viral replication. Replicon-based screens that incorporate quantifiable reporters have also been very useful for testing under reduced biosafety level conditions, although modifications to the viral genome typically need to be introduced to reduce replicon cytotoxicity (Pohjala et al. 2011; Tamm et al. 2008; Frolov et al. 1999; Dryga et al. 1997). While not always straightforward for HTS development and implementation, quantification of virus production as plaque reduction can be accomplished by staining of infected cells or through quantification of produced infectious virus in the supernatants of infected cells. Since they typically require more involved techniques, these types of assays are often used to validate hits and determine EC_{50} values.

Other phenotypic/molecular target-based screens have been designed to measure compound effects on functions of CHIKV proteins, molecular pathways, or disease-related outcomes. These types of approaches generally start with target identification, which is based on knowledge of a known protein or pathway function, and requires the ability to design assays around the desired targets. A well-described phenotypic screen for CHIKV is based on transcriptional shutoff induced by CHIKV nsP2. In this screen, CHIKV nsP2-induced host transcriptional shutoff is connected to various luciferase-based reporter gene assays, including a *trans*-reporter system that employs a Gal4 DNA-binding domain fused to Fos transcription factor (Bhat et al. 2019). This has been adapted to the HTS platform and used to identify many compounds that block nsP2 mediated host shutoff (Lucas-Hourani et al. 2013b).

A rapidly growing area in antiviral development is the development of in silico virtual screens. These types of screens are knowledge-driven approaches that combine structural information about a viral or host protein of interest for target-based screens. In silico computer-based screening of bioactive ligands ranks molecules based on their likelihood of having affinity for a certain target through a multistep process of virtual docking, scoring, validation, and simulation steps (Ekins et al. 2007a, b). Compounds identified by this type of screening method can be further optimized in silico using computer-based structural models through an understanding of the compound's structure–activity relationship (Bhat et al. 2019).

Any compound hits that come out of this type of screen should be further validated by empirical methods. Co-crystallization, differential scanning fluorimetry, and isothermal titration calorimetry are just a few. Most of these techniques require highly purified proteins. Because high-resolution structures of many CHIKV nsPs and glycoproteins are solved, they are ideal targets for the development of virtual screens and computer-aided design. Further into this chapter, we present examples of compounds identified through in silico screening methods and the methods used to validate their in vivo efficacy.

4 Small Molecule Inhibitor Strategies for Targeting CHIKV

4.1 Inhibitors of CHIKV Entry

Attachment of CHIKV particles to a host cell involves the binding of the viral E2 glycoprotein with a host cell receptor protein. CHIKV E2 has two surface-exposed domains (known as domains A and B) that are capable of binding to cells (Cho et al. 2008; Voss et al. 2010). Multiple factors are thought to be involved in CHIKV attachment including numerous proposed entry receptors including prohibitin-1, TIM-1, glycosaminoglycans, and others; thus far most of these proposed receptors act mainly as attachment factors to capture virus and facilitate entry (Hoornweg et al. 2016; Moller-Tank et al. 2013; Silva et al. 2014; van Duijl-Richter et al. 2015; Wintachai et al. 2012). Mapping studies suggest E2 domain A contains a charged heparin sulfate-binding groove that may overlap with other cellular attachment protein-binding sites (Sahoo and Chowdary 2019). Most recently, a genome-wide CRISPR-Cas9-based host gene deletion screen identified the cell adhesion molecule Mxra8 as an entry mediator for CHIKV and other alphaviruses (Zhang et al. 2018). Mxra8 binds a surface-exposed region across the A and B domains of E2, which are also speculated attachment sites (Zhang et al. 2018). Studies with Mxra8 and other potential CHIKV entry receptors presented similar conclusions that no single receptor/factor is critical for CHIKV attachment/entry, and therefore, it is a combination of host factors that likely contribute to cell attachment and entry (Moller-Tank et al. 2013; Silva et al. 2014; van Duijl-Richter et al. 2015; Wintachai et al. 2012; Zhang et al. 2018). This makes inhibition of viral entry difficult because blocking one factor of CHIKV entry will likely not be adequate to inhibit CHIKV replication. One place to start is the development of small molecule screens that target the Mxra8-binding region of E2 and the interaction sites of other putative CHIKV attachment/entry components. Treatment of mice with either an anti-*Mxra8* antibody or a *Mxra8*-Fc fusion protein reduced CHIKV infection and associated foot swelling, suggesting that Mxra8-specific small molecule inhibitors may be an effective strategy to inhibit CHIKV infection. This strategy has yet to be successfully exploited against CHIKV, and the combination of multiple entry process inhibitors is one way to inhibit CHIKV infection during the early stages.

After attachment, CHIKV is internalized largely through clathrin-mediated endocytosis (CME), although clathrin-independent entry has also been reported (Ooi et al. 2013; Bernard et al. 2010; van Duijl-Richter et al. 2015). Although micropinocytosis has been identified as a major route of entry in muscle cells, which was inhibited by 5-(*N*-ethyl-*N*-isopropyl)amiloride (EIPA) (Lee et al. 2019). Inside the early acidic endosome, the low pH environment triggers CHIKV E1-mediated fusion of the viral envelope and endosomal membranes in a process that requires the presence of cholesterol (van Duijl-Richter et al. 2015). Membrane fusion is followed by virus disassembly and release of the nucleocapsid, uncoating, and release of the viral genome into the cytosol. Interfering with the formation of clathrin-coated pits and membrane fusion by adjusting endosome pH are successful strategies to inhibit CHIKV entry.

In 2012, the first CHIKV receptor prohibitin-1 (PHB1) was described, acting as a CHIKV-binding factor in human microglia cells (Wintachai et al. 2012). In addition, a class of naturally occurring plant compounds, flavaglines (FL), as well as their synthetic analogs (FL3 and FL23), and sulfonyl amides, are able to interact with PHB1 (Chang et al. 2011; Ribeiro Morais et al. 2011), representing a class of potential CHIKV inhibitors. Replication of CHIKV in HEK293 cells was maximally inhibited when cells were treated with the synthetic flavaglines FL3 and FL23 prior to CHIKV infection (Wintachai et al. 2015). However, little inhibitor activity was detected following a post-entry treatment regimen, suggesting FL3 and FL23 are capable of only inhibiting virus entry. In PHB-CHIKV E2 co-localization, there was a significant reduction in binding interactions between PHB1 and CHIKV E2 glycoprotein in the presence of FL3 or FL23, with reduction also occurring in the presence of the sulfonyl amide 1 m inhibitor. Approximately fifty percent of treated cells still displayed signs of infection, again indicating the role of additional coreceptors or CHIKV entry mechanisms.

Epigallocatechin-3-gallate (EGCG) is another naturally occurring compound that inhibited CHIKV infection (Weber et al. 2015). A major component of green tea, EGCG has displayed antiviral abilities through an interaction with viral surface proteins that inhibits cellular attachment (Kaihatsu et al. 2018). While concurrent administration of EGCG with infectious CHIKV was able to reduce the rate of infection, this was not observed when EGCG was added following infection with CHIKV (Lu et al. 2017). This finding supports the proposed mechanism that EGCG functions as an antiviral through the inhibition of entry rather than inhibiting viral replication (Weber et al. 2015). This inhibition has been documented to affect the infectivity of both sialic acid and heparan-sulfate-binding viruses, inhibiting attachment of adenovirus, vesicular stomatitis virus, and vaccinia virus among others (Colpitts and Schang 2014).

Chloroquine is another broadly acting antiviral compound with efficacy against CHIKV. Initially developed as an antimalarial drug, chloroquine acts as a broad-spectrum antiviral through disruption of endosomal entry of viruses and inhibiting replication. Along with these activities, accumulation of chloroquine in lymphocytes and macrophages disrupts the secretion of proinflammatory molecules, such as tumor necrosis factor α (TNFα) and the receptor for TNFα

(Savarino et al. 2003). When tested for CHIKV inhibition, chloroquine was found to act in a time- and dose-specific manner. Pre-treatment inhibited viral binding to cells, likely by altering cell–virus interactions. Chloroquine was previously identified to alter the terminal glycosylation of angiotensin I converting enzyme 2 (ACE2), a receptor for severe acute respiratory syndrome coronavirus (SARS-CoV). It is likely that chloroquine may act in a similar function to disrupt cellular receptors important for CHIKV infection (Khan et al. 2010). When chloroquine was provided concurrently with infectious CHIKV, or up to 1-hour post-infection, CHIKV infection was reduced, likely through the pH modulating effects of chloroquine on endosomes. This alteration in endosomal pH presumably inhibited virus–endosome interactions required for conformational changes in unpackaging (Khan et al. 2010). While the effects of chloroquine are quite dramatic in vitro, treatment efficacy in non-human primates and humans has been limited (Roques et al. 2018).

Arbidol or umifenovir is marketed as a prophylactic antiviral treatment against influenza A and B in both Russia and China but inhibits a wide breadth of viruses including Ebola, hepatitis C, and Tacaribe virus. Maximum inhibition of CHIKV occurred when arbidol was added prior to infection with IC_{50} values ranging from 5 to 10 μg/mL (Delogu et al. 2011). CHIKV resistance selected against arbidol occurs following a single amino acid substitution at G407R of the viral E2 glycoprotein. Mechanistically, the effects of arbidol are hypothesized to occur through disrupting the formation and integrity of cytopathic vacuoles attached to endosomes and lysosome membranes due to arbidol incorporation into these membranes (Blaising et al. 2014). Derivatives of arbidol with increased potency and selectivity index have been synthesized (Di Mola et al. 2014), but additional studies are required to elucidate the mechanism of action for arbidol against CHIKV and to further develop analogs that can extend the antiviral treatment window.

Suramin is a multifunctional polysulfonated small molecule with antiviral, -neoplastic, and -nematodal activities and is currently FDA approved for the treatment of trypanosomiasis in humans. Suramin has antiviral activities against a wide variety of viruses including CHIKV (Ho et al. 2015). While multiple mechanisms for suramin activity against CHIKV have been proposed, the compound inhibits viral entry (Albulescu et al. 2015). Interestingly, molecular docking studies indicate that suramin may embed itself into the cavity present in the E1/E2 heterodimer and interfere with their function (Ho et al. 2015). Other lead compounds have been identified using computational docking using the structure for the viral envelop, but testing for antiviral activity of these leads is still pending (Agarwal et al. 2019).

CHIKV fusion and budding are influenced by the lipid composition of the viral envelope and the host cell membrane. This is particularly the case for levels and composition of cholesterol and sphingolipids. Therefore, abnormalities in lipid metabolism can affect CHIKV infection outcomes. For example, depletion of cholesterol with methyl-β-cyclodextrin prior to CHIKV infection of cells reduces infection by up to 63% (Bernard et al. 2010). Treatment of human foreskin fibroblasts with either of two cholesterol trafficking inhibitors, U18666A and imipramine, results in a dose-dependent accumulation of intracellular cholesterol and inhibition of CHIKV replication (Wichit et al. 2017). Imipramine was demonstrated to inhibit the CHIKV entry/fusion step and impair post-fusion viral RNA replication, suggesting the compound is able to interfere with two different stages of the CHIKV infection process. These results suggest other cholesterol inhibitors may have potential antiviral activities against CHIKV.

4.2 Inhibitors of CHIKV Structural Proteins

The outer envelope surface of CHIKV is made up of 80 trimeric spikes created by heterodimers of the E1 and E2 glycoproteins (Voss et al. 2010). E2 facilitates the binding of CHIKV to the surface of the host cell receptors (Weber et al. 2017) and contains a cytoplasmic tail that interacts directly with the viral capsid proteins (Mukhopadhyay et al. 2006). Upon entry, E1, a class II viral fusion protein, mediates fusion of the viral envelope and host cell endosomal membranes (Kielian et al. 2010; Uchime et al. 2013). E3, the third glycoprotein, binds exclusively to E2 forming pE2 (Li et al. 2010). E3 is cleaved from E2 by the host protease furin (Ozden et al. 2008) exposing a *N*-terminal signal peptide that targets the structural polyprotein toward the ER for initial processing (Strauss and Strauss 1994). E3 E1-E2 dimers can then form in the trans-Golgi network, with proper folding mediated by E3 (Metz and Pijlman 2016a; Wong and Chu 2018). After cleavage, E3 remains non-covalently linked to the dimers until the neutral pH of the plasma membrane causes its dissociation. This conformational change exposes the acid-sensitive region between E1 and E2, priming E1 for activation when exposed to low pH (Metz and Pijlman 2016a, b; Uchime et al. 2013).

CHIKV capsid proteins have three domains including: (1) a highly basic region that mediates non-specific RNA interactions while containing nuclear localization and export sequences; (2) a viral genomic RNA-binding region that also promotes capsid oligomerization; and (3) a serine protease capable of *cis* and *trans* cleavage of the viral capsid proteins (Metz and Pijlman 2016a; Weiss et al. 1994; Linger et al. 2004; Sokoloski et al. 2017; Aliperti and Schlesinger 1978; Choi et al. 1991; Melancon and Garoff 1987; Thomas et al. 2010). The capsid nuclear localization and export signals allow the protein to shuttle between the nucleus and cytoplasm (Thomas et al. 2013). Mutation of the capsid NES causes nuclear retention and blockage of the nuclear import system, whereas mutation of NLS attenuates the virus (Jacobs et al. 2017; Taylor et al. 2017; Thomas et al. 2013). A hydrophobic domain present in the capsid also directly interacts with the C-terminal tail of E2 to promote assembly and to facilitate budding of CHIKV through the host cell plasma membrane (Sharma et al. 2018). The capsid hydrophobic-binding pocket binds the proline-405 residue of E2, a highly conserved residue in alphaviruses (Aggarwal et al. 2012; Kim et al. 2005). Picolinic acid (PCA), a potent inhibitor of CHIKV, closely resembles the molecular structure of proline. The compound is capable of binding to the hydrophobic pocket of CHIKV capsid (Sharma et al. 2016). Treatment with PCA results in a significant reduction in vRNA levels and infectious virus, suggesting possible inhibitory effects on viral disassembly, replication, or nucleocapsid assembly (Sharma et al. 2016). Although the exact mechanism by which it functions remains unclear, the observed antiviral properties of PCA against CHIKV demonstrate the importance of blocking the hydrophobic domain of capsid. CHIKV capsid and protease activity continue to be an under-utilized target for inhibitor development with the potential to be a critical target for CHIKV inhibitors.

4.3 Inhibitors of CHIKV Non-structural Proteins

4.3.1 nsP1

The nsP1 protein is a viral mRNA capping enzyme with both methyltransferase and guanylyltransferase (GTase) activities. The protein is responsible for capping and methylation of new synthesized viral RNA protecting it from degradation by host exonucleases and directing efficient translation of viral mRNAs (Rupp et al. 2015; Wong and Chu 2018). Additionally, nsP1 plays an important role in RC formation and localization, which is directed by an alpha-helical amphipathic loop and palmitoylation site that allows nsP1 containing RCs to dock to the host cell plasma membrane. This process tethers RCs within spherules (Spuul et al. 2007). nsP1 is involved in the recruitment of other nsPs into spherules required for the formation of functional RCs (Abu Bakar and Ng 2018; Salonen et al. 2003). nsP1 is also involved in releasing budding particles from the cell membrane through interactions with the host protein tetherin (BST-2) (Jones et al. 2013). These functions of nsP1 make it an appealing target for drug discovery as blocking nsP1 functions disrupt RC formation and prevent vRNA synthesis (Abu Bakar and Ng 2018). However, the discovery and design of inhibitors targeting these functions have been difficult due to the lack of definitive structural knowledge about the interactions between nsPs and limited information about the structure of CHIKV spherules.

The most fully characterized inhibitors of CHIKV nsP1 target the GTase function of the capping machinery. MADTP-314 and MADTP-372 are CHIKV and VEEV nsP1 inhibitors that are part of the [1,2,3]trizolo[4,5-d]pyrimidin-7(6 h)-ones (MADTP) compound series identified using a cell-based CHIKV replication screen (Gigante et al. 2014, 2017). CHIKV resistance to MADTP-314 occurs through a P34S mutation in nsP1 GTase functional domain that was validated by reverse genetics. In vitro VEEV nsP1 GT assay studies determined that MADTP-372 disrupted GTase activity and downstream capping reactions, which are suggested to occur by either disruption of m^7GTP-nsP1 complex formation or inhibition of guanylation (Delang et al. 2016; Gigante et al. 2017). Recently, an HTS was developed to identify compounds that inhibit the formation of the 5' cap by measuring competition for the GTP-binding site on CHIKV nsP1 using fluorescently labeled GTP (Bullard-Feibelman et al. 2016). This method identified the natural compound lobaric acid, as a GTP competitor for nsP1 binding and inhibitor of the guanylation step of the capping reaction (Feibelman et al. 2018).

4.3.2 nsP2

CHIKV nsP2 is a large protein with at least four enzymatic functions including: (1) helicase activity by unwinding double-stranded RNA in the 5′ to 3′ direction; (2) nucleotide triphosphatase (NTPase) activity; (3) RNA 5′ triphosphatase activity; and (4) papain-like cysteine protease activity (Rupp et al. 2015; Wong and Chu 2018; Das et al. 2014b; Karpe et al. 2011; Ramakrishnan et al. 2017). Activity of the C-terminal protease domain is responsible for processing the viral nsP1234 polyprotein (Ramakrishnan et al. 2017). Additionally, a portion of nsP2 can localize to the nucleus where it plays a role in the shutoff of host transcription by mediating degradation of the RNA polymerase subunit II Rbp1 (Akhrymuk et al. 2012). nsP2 also mediates the shutoff of host translation by interacting with a number of ribosomal proteins (Strauss and Strauss 1994). Host cell transcriptional and translational shutoffs occur without any obvious negative effects to CHIKV replication. Although mutation of the nsP2 NLS of Semliki Forest virus (SFV) prevented the protein from entering the nucleus; this process also reduced SFV-induced cell death, likely due to a reduction in cytotoxicity associated with host shutoff (Tamm et al. 2008). Nuclear localized nsP2 is also capable of inhibiting innate immunity by suppression of JAK/STAT signaling. Therefore, multiple modes of action may be at play (Bhalla et al. 2016; Breakwell et al. 2007; Frolov et al. 2009; Gorchakov et al. 2005). These antiviral functions of nsP2 have earned the protein designation as a virulence factor as well. Thus, due to these effects on RNA replication and those directed against the host, nsP2 is a valid antiviral target for inhibiting CHIKV. For example, a high-throughput phenotypic screen to identify compounds that target virus-mediated host transcriptional shutoff induced by nsP2 was developed utilizing a trans-reporter where expression of a luciferase gene is driven by an artificial transcription factor (Lucas-Hourani et al. 2013b). This method successfully identified a natural compound that blocked nsP2 activity and inhibited CHIKV replication.

In recent years, the nsP2 protease function has become a major target of interest due to its essential role in viral replication and the success of FDA-approved inhibitors targeting HIV and HCV proteases (Manns and von Hahn 2013; Pokorna et al. 2009; De Clercq 2007). Peptidomimetic compounds that target nsP2 protease activity have been developed using a number of biochemical tools including a FRET-based protease assay (Singh et al. 2018). The crystal structures of CHIKV and VEEV nsP2 have been solved, and in silico screening for nsP2 protease inhibitors has begun. Millions of small molecule structures have been tested by a variety of virtual screening methods to identify compounds that interact with CHIKV nsP2 protease domain (Bassetto et al. 2013; Agarwal et al. 2015) and to explore the possible mechanism of action of compounds through in silico docking (Jadav et al. 2015). These types of in silico studies have generated a number of promising lead compounds; however, experiments that confirm antiviral activity and/or target specificity are needed to determine whether the in silico identified compounds inhibit by the predicted functions (Abu Bakar and Ng 2018; Dhindwal et al. 2017). Albeit, in silico predicted inhibitors targeting the catalytic site of the nsP2 protease were active in viral inhibition assays and nsP2 protease function assays. Some of the predicted compounds decreased nsP2 protease function, viral RNA synthesis, and release of infectious viral particles (Das et al. 2016). This study supports the feasibility of virtual screens to identify target-specific viral inhibitors.

4.3.3 nsP3

The functional role of nsP3 during CHIKV replication is still unclear. Structurally the nsP3 protein is divided into three domains: (1) an N-terminal macrodomain; (2) an alphavirus unique domain (AUD) containing a zinc-binding region; and (3) a C-terminal hypervariable domain (Abu Bakar and Ng 2018). The highly conserved macrodomain is thought to regulate CHIKV replication through the binding of RNA and ADP-ribose and ADP-ribosyl hydrolase activities (Malet et al. 2009; McPherson et al. 2017; Abraham et al. 2018). These activities confer efficient CHIKV replication, indicating a need for the virus to evade host ADP-ribosylation of proteins or RNA but how this is important is still unclear. The highly conserved nature of the macrodomain and residues of the ADP-ribose-binding pocket within the alphavirus family and other RNA viruses makes it an ideal site for further development of macrodomain-specific antivirals. Using computer-aided design, we have screened a small fragment compound library to identify small molecules that bind within the ADP-ribose-binding pocket of the CHIKV macrodomain. We validated one fragment through crystal soaking and NMR. However, the small size of the fragments prohibits their utility as inhibitors in cell-based assays. We are currently utilizing computer-aided design to build larger fragments that increase their specificity and activity as CHIKV inhibitors. Other in silico studies aimed at similarly identifying small molecule inhibitors of the macrodomain ADP binding identified both naturally occurring small molecules (flavonoids) and pharmaceutical compounds with the potential to bind CHIKV nsP3 including the flavonoid naringenin (Nguyen et al. 2014; Pohjala et al. 2011; Seyedi et al. 2016).

Recent mutational studies of the CHIKV AUD suggest this domain determines species specificity and plays a key role in virus genome and RNA transcription assembly (Gao et al. 2019). Mutation of the AUD impaired subgenomic RNA synthesis, RNA-binding activity of the domain, and subcellular localization of nsP3 during CHIKV replication, in turn reducing virus production. This analysis highlights the potential for the nsP3 AUD as an antiviral target. nsP3 proteins from other alphaviruses reportedly interact with a diverse range of host factors including sphingosine kinase 2 (SK2), Hsp90B, PI3K-AKT-mTOR pathway, and IkkB with mostly proviral outcomes (Lark et al. 2017), further supporting future investigations into their utility as antiviral targets by disruption of these virus–host protein interactions necessary to enhance virus replication. The C-terminal domain promotes CHIKV replication by interfering with stress granule formation through interactions with GTPase-activating protein (SH3 domain)-binding protein 1 (G3BP1) (Panas et al. 2012, 2014; Fros et al. 2012) and the mosquito version of G3BP1 called Rasputin (Fros et al. 2015). The heat shock protein Hsp-90 has also been shown to interact with nsP3, but the role in virus replication is still unclear (Rathore et al. 2014). Despite the many potential target sites and key host and viral interactions of nsP3, only a limited number of studies have focused on developing antivirals targeting nsP3.

4.3.4 nsP4

The alphavirus RNA-dependent RNA polymerase (RdRp) is encoded within nsP4, and this protein is the most highly conserved in the alphavirus family (Rupp et al. 2015; Weston et al. 2005; Pietila et al. 2017). Alphavirus RdRps are responsible for the replication of the viral genome and subgenomic transcripts. As part of the P123 early replication complex, nsP4 mediates the synthesis of the negative genomic strand from the incoming genome (Pietila et al. 2017). Once fully active, the RC complex containing the polymerase shifts to the synthesis of the 49S genomic RNA and 26S subgenomic RNA from these negative strand templates. nsP4 N-terminal domain also contains adenylyltransferase (TATase) activity, which was identified through mutational studies that indicated a role in adding or maintaining the 3' poly-A tail at the end of the genome (Rubach et al. 2009; Tomar et al. 2006). The N-terminal domain of nsP4 also contains an alphavirus-specific domain that is important for the interaction with the P123 complex and formation of RCs. Deletion of the 97 N-terminal residues prevents de novo RdRp activity regardless of the presence of P123 as well as the association with the P123 complex. This finding suggests that this region of nsP4 may be a valid target for small molecule inhibitor development (Rubach et al. 2009; Tomar et al. 2006). nsP4 also interacts with the host protein Hsp90α via an unknown mechanism and, as mentioned above, inhibition of Hsp90α decreases viral RNA and protein synthesis (Rathore et al. 2014).

Due to their unique structure, RNA virus RdRp is an important target for drug development. As such a number of compounds have been developed that block RNA virus replication. Initially developed as an anti-influenza virus inhibitor, favipiravir (T-705) has also shown good activity against many other divergent RNA virus polymerases including those encoded by alphaviruses (Furuta et al. 2017). However, T-705 is not active against DNA or DNA-dependent RNA polymerases making it selective to both plus and minus strand RNA viruses. Favipiravir is quickly converted into the triphosphate active form within cells, which is recognized as a substrate for the viral RdRp (Furuta et al. 2005). Functionally T-705 competitively inhibits the incorporation of ATP and GTP by the RdRp leading to chain termination (Delang et al. 2014; Furuta et al. 2013). T-1105 is a T-705 analog that effectively inhibits CHIKV nsP4 through interactions with the Lys-291 residue. T-705 is active against a wide variety of alphaviruses and other RNA viruses, potentially because the Lys-291 residue is conserved in the polymerases of positive-sense RNA viruses (Delang et al. 2014). β-D-N^4—hydroxycytidine (NHC) is another nucleoside analog that targets viral RdRps. These types of nucleosides tend to work potently and for the development of even low-level resistance against NHC, the alphavirus VEEV requires the acquisition of multiple cooperative mutations within the RdRp domain of nsP4 (Urakova et al. 2018). The FDA-approved drug Sofosbuvir is a UMP prodrug that gets converted to the active form in cells where it acts as a chain terminator for flavivirus RNA polymerases. Sofosbuvir was validated to also bind CHIKV nsP4 and inhibit RNA synthesis and virus replication in cultured cells and in vivo (Ferreira et al. 2019). NHC is also capable of inhibiting CHIKV replication (Ehteshami et al. 2017). In addition to these nucleoside analogs, HTS of chemical compound libraries identified a non-nucleoside benzimidazole compound possessing inhibitory activity against nsP4 (Wada et al. 2017). This compound inhibited the RdRp function of nsP4 by targeting residue Met-2295, potentially inhibiting the RdRp's ribonucleotide selection function (Wada et al. 2017). The potential is high for developing other nucleoside analogs or compounds that target alphavirus nsP4.

4.4 Inhibitors of CHIKV RNA Genome Replication

During replication, the viral RNA genome is first converted to the minus strand utilizing the P123/nsP4 complex. Once fully processed, the replication complex then mediates synthesis of the full-length genomic RNA (plus strand) as well as the subgenomic RNA. Targeting of many different nsP functions, as described above, can have pleiotropic effects on viral replication including vRNA synthesis. However, there are also a number of broad-spectrum RNA genome replication inhibitors described with activity against CHIKV. One of the first to be described is Ribavirin, which is a broad-spectrum antiviral that has already been approved for the treatment of RSV in infants and chronic hepatitis C infections (Pawlotsky 2014; Turner et al. 2014). Ribavirin has demonstrated antiviral activity against CHIKV and exhibited synergism with both doxycycline and IFN-α (Briolant et al. 2004; Rothan et al. 2015). Mechanistically, ribavirin is a guanosine analog, with the major proposed biomechanisms for inhibition of RNA viruses including interference with inosine monophosphate dehydrogenase (IMPDH) function leading to depletion of GTP pools (Leyssen et al. 2006), as well as inhibition of viral RNA capping (Paeshuyse et al. 2011). Other possible mechanisms include an increased mutation rate as a result of the incorporation of ribavirin by the RdRp (Paeshuyse et al. 2011). Additional studies to determine the mechanism leading to ribavirin inhibition of CHIKV are ongoing, but it is important to note that it is effective only during the early stages of the CHIKV replication cycle (Mishra et al. 2016).

RNA viruses are highly mutable making them capable of developing resistance to small molecules. However, due to this relatively high mutation rate most RNA virus polymerases are unable to detect or repair damaged or altered nucleotides. This fact allows one to design nucleotide analogs as either chain terminators or ones that increase the mutation frequency above tolerable rates so that the virus becomes genetically unstable. A broad-spectrum viral genome replication inhibitor is the uridine analog 6-azauridine. Compared to ribavirin, 6-azauridine is more effective against CHIKV in infected cells (Briolant et al. 2004; Pohjala et al. 2011). Similar to other nucleoside analogs, 6-azauridine most likely interferes with cellular UTP metabolism and the nucleoside analog incorporates into CHIKV RNA leading to genome error catastrophe (Rada and Dragun 1977; Scholte et al. 2013). While 6-azauridine has been approved for clinical use against psoriasis, further testing in animal models of viral infection is needed to examine in vivo antiviral activity (Deneau and Farber 1975; Crutcher and Moschella 1975).

5 Targeting Host Factors Involved in CHIKV Replication

The ability of viruses to rapidly evolve and select for resistant mutations to single antiviral treatments makes it necessary to identify antiviral drug cocktails containing individual compounds that lack overlapping resistance markers to increase therapeutic efficacy. Another way to reduce antiviral resistance is to target host proteins or processes required for virus replication. Development of host-targeting antivirals has another advantage in that they may have increased breadth of antiviral activity for viruses that share a cellular pathway. For example, harringtonine and its analogs homoharringtonine and cephalotaxine alkaloids were identified using an immunofluorescence-based screen of small molecule inhibitors derived from natural products to have potent anti-CHIKV activity (Kaur et al. 2013). Harringtonine functions as an inhibitor of eukaryotic translation by blocking the large ribosomal subunit (Fresno et al. 1977). The compound also inhibits translation of Epstein–Barr virus and influenza virus and has been used for translational profiling experiments (Bencun et al. 2018; Machkovech et al. 2019). Harringtonine blocked translation of CHIKV nsPs and inhibited viral RNA synthesis and subsequent production of structural proteins, indicating that the compound inhibits early viral translation events (Kaur et al. 2013). Resistance to harringtonine has yet to be reported for any virus. It is important to keep in mind the potential for off-target effects and isoform specificity that can limit host-directed antivirals especially for in vivo use.

Different screening methods have been used to identify CHIKV/host interactions important for the virus lifecycle. Linking the results from a human whole genome-wide loss-of-function siRNA screen with a drug repurposing database search was used to rapidly identify small molecule inhibitors of CHIKV (Karlas et al. 2016). In the CHIKV/HEK-293 cell loss-of-function screen, knockdown of 156 host genes increased virus replication, whereas 41 displayed antiviral activity. To identify potential antiviral small molecule inhibitors, the validated proviral factors were used to screen available databases that link their drugs to their experimentally validated target proteins. Using this process, 20 compounds were identified that interact with gene products of 14 CHIKV proviral factors that span six unique pathways including vacuolar-type H+ ATPase, CDD-like kinase 1 (CLK1), fms-related tyrosine kinase 4 (FLT4) calmodulin signaling, fatty acid synthesis, and lysine acetyltransferase 5 (KAT5). All 20 compounds inhibit CHIKV replication but, as expected, some display cytotoxicity. A combination regimen containing TOFA, a fatty acid synthesis inhibitor, and the calmodulin inhibitor pimozide showed increased inhibition of CHIKV and significantly reduced CHIKV-induced footpad swelling (Karlas et al. 2016). As discussed below, research into host factors involved in CHIKV replication has identified additional potential inhibitors of virus replication and disease.

5.1 *Protease Inhibitors*

Host proteases are a diverse group of enzymes that catalyze the cleavage of the same or other proteins. The three large groupings of proteases include serine, cysteine, and metalloproteinases. The proteases recognize specific substrate amino acid sequences and perform cleavage of the scissile bond. Host furin is a serine-like protease that processes a number of different substrates including host-derived proalbumin and transforming growth factor beta. However, as a resident of the TGN, furin also cleaves a number of viral glycoproteins including tick-borne encephalitis virus, HIV gp160, and HCMV gB; and this process typically activates these proteins (Hallenberger et al. 1992; Stangherlin et al. 2017). Similarly, furin is involved in processing of the alphavirus envelope pE2/E3 precursor at short multibasic motifs during virion transport through the TGN (Klimstra et al. 1999; Ozden et al. 2008). This processing event is required for the formation of infectious particles for many alphaviruses including CHIKV, which validates furin as an antiviral target (Heidner et al. 1994, 1996; Klimstra et al. 1999). The irreversible furin inhibiting peptide decanoyl-RVKR-chloromethyl-ketone (Dec-RVKR-cmk) significantly inhibits CHIKV infection in human muscle satellite cells by impairing the formation of mature virus particles (Ozden et al. 2008). Interestingly, therapy combining dec-RVKR-cmk with chloroquine had additive effects resulting in a near-complete suppression of virus spread and yield when added just prior to CHIKV infection. Obvious issues with selectivity and toxicity related to the plethora of cellular furin cleavage substrates as well as the large size of current furin inhibitors will need to be considered when assessing the antiviral therapeutic value for inhibitors of furin (Subudhi et al. 2018).

5.2 *Pyrimidine and Purine Synthesis Inhibitors*

RNA viruses rely heavily on the host pool of nucleosides for efficient replication. Changes in the concentration of ribonucleotide triphosphate pools can influence the ability of the RNA virus to replicate and possibly increase mutation frequency (Ortiz-Riano et al. 2014). A number of HTS campaigns have identified inhibitors of pyrimidine synthesis as potent antivirals with broad activity (Lucas-Hourani et al. 2013a; Chung et al. 2016; Hoffmann et al. 2011; Smee et al. 2012; Wang et al. 2011). The small molecule DD264 was identified through a cell-based HTS assay to identify molecules that stimulate the interferon response, and through mechanism of action, experiments were identified as an inhibitor of de novo pyrimidine synthesis suggesting a unique link between the interferon response and pyrimidine biosynthesis (Lucas-Hourani et al. 2013a). DD264 inhibition of virus replication was dependent upon the activation of IRF1 suggesting an important role in innate immune activation. Interestingly, DD264 inhibition of CHIKV was blocked when the pyrimidine uridine but not purine guanosine was added to the culture medium also supporting that lowered pyrimidine levels are responsible for the activity of DD264. Dihydroorotate dehydrogenase (DHODH), the fourth enzyme in the pyrimidine biosynthetic pathway, was identified as the target of DD264. DD264 has proven to be a useful tool to better understand the link between the innate immune response and pyrimidine biosynthesis during CHIKV replication. Targeting the mitochondrial electron transport with antimycin A also inhibits de novo pyrimidine synthesis resulting in a broad-spectrum antiviral effect (Raveh et al. 2013). Compounds that target purine biosynthesis have similar antiviral properties against CHIKV. Mycophenolic acid (MPA) inhibits the cellular enzyme inosine monophosphate dehydrogenase that is required for guanine synthesis. MPA inhibits CHIKV replication by blocking viral genome synthesis (Khan et al. 2011). The in vivo therapeutic application of pyrimidine biosynthesis inhibitors is complicated by the high uridine concentration in the body that can negate the antiviral effects, but there may be a utility as site-specific antiviral treatments.

5.3 Cellular Kinase Inhibitors

Viruses modify host kinase signaling pathways in order to adjust the host environment to promote their replication (Keating and Striker 2012). The PI3K-AKT-mTOR pathway is involved in cell survival and alphaviruses activate this pathway. Semliki Forest virus nsP3 directly activates AKT at the plasma membrane where it is probably involved in the formation of the replication complex (Spuul et al. 2010). However, this effect might be virus-specific since CHIKV replication complex formation was not dependent upon this pathway (Thaa et al. 2015). CHIKV nsP3 is also involved in recruiting sphingosine kinase to the replication complex (Reid et al. 2015). There are a number of high-throughput methods that have been used to identify additional host kinases and signaling pathways involved in CHIKV replication and potential inhibitors. HTS screens utilizing kinase inhibitor libraries such as the BioFocus kinase inhibitor library identified six lead hit compounds with the most potent compound CND3514 a thiozole-4-carboxaminde core scaffold inhibitor with an $EC_{50} = 2.2\ \mu M$ (Cruz et al. 2013). Other approaches including genome wide or kinase focused siRNA library screens have also been used (Reid et al. 2015). The use of multiplexed inhibitor beads to profile changes in the kinome has also been employed to identify kinases relevant to CHIKV replication (Broeckel et al. 2019). Through this process, changes in the abundance/activity of the Src family kinase (SFK)-

phosphatidylinositol 3-kinase (PI3K)-AKT-mTORC signaling pathway during the course of CHIKV were discovered. Inhibition of this pathway with the SFK inhibitor dasatinib blocked replication of CHIKV and multiple other alphaviruses in human fibroblasts. In mechanism of action studies, dasatinib was found to block CHIKV subgenomic RNA translation, significantly reducing structural protein levels, without affecting synthesis of viral genomic or subgenomic RNA (Broeckel et al. 2019). A similar effect was observed with the mTORC1/2 inhibitor Torin. These results were in part due to a decreased amount of CHIKV RNA associated with polysomes during replication, suggesting CHIKV relies on SFKs for structural protein synthesis (Broeckel et al. 2019).

Protein kinase C (PKC) is a serine/threonine kinase that is recruited to the plasma membrane upon cellular activation in response to a number of stimuli. PKC may play a role in early viral entry steps that involve endosomal trafficking. Inhibition of PKC with H-7 blocked entry for a number of enveloped viruses including Sindbis virus (Constantinescu et al. 1991). PKC modulators have also been tested for their ability to inhibit CHIKV replication. Prostratin and 12-O-tetradecanoylphorbol-13-acetate (TPA) inhibited CHIKV replication in Vero cells but TPA may be CHIKV-specific as it fails to block other alphaviruses (Bourjot et al. 2012). Aplysiatoxin analogs, known PKC activators, debromoaplysiatoxin and 3-methoxydebromoaplysiatoxin were also reported to inhibit CHIKV (Gupta et al. 2014). The pan-PKC modulator byrostatin 21 also potently inhibited CHIKV replication without modulating the cellular PKC pathways, which suggest that these compounds may also work through PKC-independent pathways. (Staveness et al. 2016). Deciphering the mechanisms of how PKC modulates the CHIKV lifecycle requires further investigation. Similar to the other host targeted antivirals discussed, the clinical use of PKC modulators will be limited because of the importance of PKC in normal cell survival. Additional studies are required to fully elucidate the specific cellular kinase pathways involved in CHIKV replication in order to fully explore the development of novel small molecule antivirals.

5.4 *Inhibitors of Protein Chaperones*

Cytoplasmic proteins and those traversing through the cellular secretory compartment require chaperones for proper folding and disulfide bond formation that is important for proper trafficking, increased stability, and improved function. Two categories of chaperones have been shown to be involved in CHIKV replication including the Heat shock protein-90 (Hsp-90) and protein disulfide isomerases (PDI). For example, Hsp-90 is a highly abundant chaperone that is utilized by both cellular and viral proteins to ensure proper folding, maturation, localization, and turnover of substrate proteins. Hsp-90 plays an important role in the replication of many RNA and DNA viruses making it a possible target for broad-spectrum antiviral development. The Hsp-90 inhibitor geldanamycin and synthetic analogs of geldanamycin HS-10 and SNX-2112 all inhibit CHIKV replication

(Rathore et al. 2014). Geldanamycin inhibition has shown that Hsp-90 is essential during the early stages of CHIKV replication by promoting nsP2 stability (Das et al. 2014a). Hsp-90 also interacts with CHIKV nsP3 and nsP4, and Hsp-90α may play an important role in the stabilization of nsP4 and formation of the replication complex (Rathore et al. 2014). HS-10 and SNX-2112 treatment significantly reduced serum viral titers at 48 hpi and decreased CHIKV-induced joint swelling disease and inflammatory cytokine production in SVA129 infected mice (Rathore et al. 2013). However, the development of Hsp-90 inhibitors for in vivo use has been difficult because the chaperone is involved in many cellular processes and signaling pathways that can cause a certain level of cytotoxicity when inhibited. A recently developed second-generation Hsp-90 inhibitor called Ganetespib has decreased cytotoxicity and in vivo safety (Jhaveri and Modi 2015).

Inhibitors of cellular PDIs block CHIKV infection by decreasing the infectivity per particle ratio of secreted viruses (Langsjoen et al. 2017). Since CHIKV E1 and E2 glycoproteins require specific disulfide bonding patterns between conserved cysteine residues, the reduction in infectivity is likely to require host PDI for envelope protein function. Consistent with this hypothesis, PDI inhibitors decreased cell–cell fusion events facilitated by E1. Auranofin, an FDA-approved thioredoxin reductase (TRX-R) inhibitor, was the most promising compound with a therapeutic index of 104.5 at 12 hpi, and was efficacious in mouse models of CHIKV infection and disease (Langsjoen et al. 2017).

6 Conclusions

The development of inhibitors against Chikungunya virus is critical for treating infected patients to prevent or reduce transmission and disease. CHIKV remains a clinically relevant human pathogen due to the severity and chronicity of disease and explosive nature of viral epidemics. Recent development of inhibitors against CHIKV has identified a number of viable viral and cellular targets that, when blocked, can robustly inhibit virus replication. However, there are a number of challenges that remain to overcome in order to successfully treat patients. For example, the virus rapidly mutates indicating that the virus can quickly develop resistance to most single-drug regimens. This would indicate that two or more drugs that target unique aspects of virus replication are required to limit resistance that would render the drug ineffective. Another major issue that needs to be addressed is the aspect of determining what the therapeutic window is for treating CHIKV infection and disease. Due to the chronicity of the viral infection and disease, when testing new antivirals, it is important to not only assess efficacy during the acute phase but also the chronic phase in order to establish the effective therapeutic window. The process of developing inhibitors and identifying the antiviral targets will continue to also improve our understanding of the virus lifecycle (Table 1).

Table 1 Viral and host-directed antivirals against Chikungunya virus

Compound	Target/MOA	Validation method	References
Inhibitors of CHIKV entry			
FL23/FL3	Entry	In vitro	Wintachai et al. (2015)
EGCG	Entry	In vitro	Weber et al. (2015)
Chloroquine	Entry	In vitro	Khan et al. (2010)
Arbidol/Emifenovir	Entry	In vitro	Delogu et al. (2011)
IIIe 7/IIIf (Arbidol derivatives)	Entry	In vitro	Di Mola et al. (2014)
Imipramine	Entry	In vitro	Wichit et al. (2017)
U18666A	Entry	In vitro	Wichit et al. (2017)
Suramin	Entry	In vitro	Ho et al. (2015)
Inhibitors of CHIKV structural proteins			
Picolinic acid	Capsid protein	In vitro	Sharma et al. (2016)
Inhibitors of CHIKV non-structural proteins			
nsP1			
MADTP	nsP1	In vitro	Delang et al. (2016), Gigante et al. (2014, 2017)
Lobaric acid	nsP1	In vitro	Feibelman et al. (2018)
nsP2			
2E	nsP2	In silico/ in vitro	Bassetto et al. (2013), Das et al. (2016)
IDI452-2	nsp2	In vitro	Lucas-Hourani et al. (2013b)
NCL1610	nsP2	In silico	Nguyen et al. (2015)
ZINC67680487	nsP2	In silico	Jadav et al. (2015)
ZINV0472520	nsP2	In silico/ in vitro	Jadav et al. (2017)
nsP3			
Naringenin	nsp3	In silico/ in vitro	Pohjala et al. (2011), Seyedi et al. (2016)
NCA_25457	nsp3	In silico	Nguyen et al. (2014)
NC_345647	nsp3	In silico	Nguyen et al. (2014)
nsP4			
Favipiravir (T705)	nsP4	In vitro/ in vivo	Delang et al. (2014)
B NHC	nsP4	In vitro	Ehteshami et al. (2017)
Compound A	nsP4	In vitro	Wada et al. (2017)

(continued)

Table 1 (continued)

Compound	Target/MOA	Validation method	References
Inhibitors of CHIKV RNA genome replication			
Ribavirin	Nucleoside analog	In vitro	Briolant et al. (2004), Pohjala et al. (2011)
6-Azauridine	Nucleoside analog	In vitro	Briolant et al. (2004), Pohjala et al. (2011)
Host-targeting compounds			
Pimozide	Calmodulin	In vitro/ in vivo	Karlas et al. (2016)
TOFA	Fatty acid synthesis	In vitro/ in vivo	Karlas et al. (2016)
Dec-RVKR-cmk	Furin	In vitro	Ozden et al. (2008)
Pyrimidine and purine synthesis inhibitors			
DD264	De novo pyrimidine biosynthesis	In vitro	Lucas-Hourani et al. (2013a)
Mycophenolic acid	Guanine synthesis	In vitro	Khan et al. (2011)
Inhibitors of cellular kinases			
CND3514	Kinase	In vitro	Cruz et al. (2013)
Dasatinib	Src family kinases	In vitro	Broeckel et al. (2019)
Torin	mTORC 1/2	In vitro	Broeckel et al. (2019)
Prostratin	PKC	In vitro	Bourjot et al. (2012)
12-O-tetradecanoylphorbol-13-acetate	PKC	In vitro	Bourjot et al. (2012)
Debromoaplysiatoxin	PKC	In vitro	Gupta et al. (2014)
3-Methoxydebromoaplysiatoxin	PKC	In vitro	Gupta et al. (2014)
Bryostatin-21	PKC	In vitro	Staveness et al. (2016)
Inhibitors of protein chaperones			
Geldanamycin	Hsp-90	In vitro	Rathore et al. (2014)
HS-10	Hsp-90	In vitro/ in vivo	Rathore et al. (2014)
17-AAG	Hsp-90	In vitro/ in vivo	Nayak et al. (2017)
SNX-2112	Hsp-90	In vitro/ in vivo	Rathore et al. (2014)
Auranofin	Thioredoxin reductase	In vitro/ in vivo	Langsjoen et al. (2017)
PACMA31	Protein disulfide isomerase	In vitro/ in vivo	Langsjoen et al. (2017)

Acknowledgements We would like to thank Andrew Townsend for graphical work included in Fig. 2 and Patricia Smith for editorial comments to the manuscript. This work was supported by a grant from the National Institutes of Health (1U19AI109680-01).

References

Abraham R, Hauer D, McPherson RL, Utt A, Kirby IT, Cohen MS et al (2018) ADP-ribosyl-binding and hydrolase activities of the alphavirus nsP3 macrodomain are critical for initiation of virus replication. Proc Natl Acad Sci USA. 115(44):E10457-E66. https://doi.org/10.1073/pnas.1812130115

Abu Bakar F, Ng LFP (2018) Nonstructural proteins of alphavirus-potential targets for drug development. Viruses 10(2). https://doi.org/10.3390/v10020071

Agarwal G, Gupta S, Gabrani R, Gupta A, Chaudhary VK, Gupta V (2019) Virtual screening of inhibitors against envelope glycoprotein of Chikungunya virus: a drug repositioning approach. Bioinformation 15(6):439–447. https://doi.org/10.6026/97320630015439

Agarwal T, Asthana S, Bissoyi A (2015) Molecular modeling and docking study to elucidate novel Chikungunya virus nsP2 protease inhibitors. Indian J Pharm Sci 77(4):453–460

Aggarwal M, Tapas S, Preeti, Siwach A, Kumar P, Kuhn RJ et al (2012) Crystal structure of aura virus capsid protease and its complex with dioxane: new insights into capsid-glycoprotein molecular contacts. PLoS One 7(12):e51288. https://doi.org/10.1371/journal.pone.0051288

Akhrymuk I, Kulemzin SV, Frolova EI (2012) Evasion of the innate immune response: the old world alphavirus nsP2 protein induces rapid degradation of Rpb1, a catalytic subunit of RNA polymerase II. J Virol 86(13):7180–7191. https://doi.org/10.1128/JVI.00541-12

Albulescu IC, van Hoolwerff M, Wolters LA, Bottaro E, Nastruzzi C, Yang SC et al (2015) Suramin inhibits Chikungunya virus replication through multiple mechanisms. Antiviral Res 121:39–46. https://doi.org/10.1016/j.antiviral.2015.06.013

Aliperti G, Schlesinger MJ (1978) Evidence for an autoprotease activity of sindbis virus capsid protein. Virology 90(2):366–369. https://doi.org/10.1016/0042-6822(78)90321-5

Bandeira AC, Campos GS, Sardi SI, Rocha VF, Rocha GC (2016) Neonatal encephalitis due to Chikungunya vertical transmission: first report in Brazil. IDCases 5:57–59. https://doi.org/10.1016/j.idcr.2016.07.008

Barton DJ, Sawicki SG, Sawicki DL (1991) Solubilization and immunoprecipitation of alphavirus replication complexes. J Virol 65(3):1496–1506

Bassetto M, De Burghgraeve T, Delang L, Massarotti A, Coluccia A, Zonta N et al (2013) Computer-aided identification, design and synthesis of a novel series of compounds with selective antiviral activity against Chikungunya virus. Antiviral Res 98(1):12–18. https://doi.org/10.1016/j.antiviral.2013.01.002

Bencun M, Klinke O, Hotz-Wagenblatt A, Klaus S, Tsai MH, Poirey R et al (2018) Translational profiling of B cells infected with the Epstein-Barr virus reveals 5′ leader ribosome recruitment through upstream open reading frames. Nucleic Acids Res 46(6):2802–2819. https://doi.org/10.1093/nar/gky129

Bernard E, Solignat M, Gay B, Chazal N, Higgs S, Devaux C et al (2010) Endocytosis of Chikungunya virus into mammalian cells: role of clathrin and early endosomal compartments. PLoS ONE 5(7):e11479. https://doi.org/10.1371/journal.pone.0011479

Bhalla N, Sun C, Metthew Lam LK, Gardner CL, Ryman KD, Klimstra WB (2016) Host translation shutoff mediated by non-structural protein 2 is a critical factor in the antiviral state resistance of Venezuelan equine encephalitis virus. Virology 496:147–165. https://doi.org/10.1016/j.virol.2016.06.005

Bhat SM, Mudgal PP, Sudheesh N, Arunkumar G (2019) Spectrum of candidate molecules against Chikungunya virus—an insight into the antiviral screening platforms. Expert Rev Anti Infect Theory 17(4):243–264. https://doi.org/10.1080/14787210.2019.1595591

Blaising J, Polyak SJ, Pecheur EI (2014) Arbidol as a broad-spectrum antiviral: an update. Antiviral Res 107:84–94. https://doi.org/10.1016/j.antiviral.2014.04.006

Borgherini G, Poubeau P, Jossaume A, Gouix A, Cotte L, Michault A et al (2008) Persistent arthralgia associated with Chikungunya virus: a study of 88 adult patients on reunion island. Clin Infect Dis 47(4):469–475

Bourjot M, Delang L, Nguyen VH, Neyts J, Gueritte F, Leyssen P et al (2012) Prostratin and 12-*O*-tetradecanoylphorbol 13-acetate are potent and selective inhibitors of Chikungunya virus replication. J Nat Prod 75(12):2183–2187. https://doi.org/10.1021/np300637t

Breakwell L, Dosenovic P, Karlsson Hedestam GB, D'Amato M, Liljestrom P, Fazakerley J et al (2007) Semliki Forest virus nonstructural protein 2 is involved in suppression of the type I interferon response. J Virol 81(16):8677–8684

Briolant S, Garin D, Scaramozzino N, Jouan A, Crance JM (2004) In vitro inhibition of Chikungunya and Semliki Forest viruses replication by antiviral compounds: synergistic effect of interferon-alpha and ribavirin combination. Antiviral Res 61(2):111–117

Broeckel R, Sarkar S, May NA, Totonchy J, Kreklywich CN, Smith P et al (2019) Src Family kinase inhibitors block translation of alphavirus Subgenomic mRNAs. Antimicrob Agents Chemother 63(4). https://doi.org/10.1128/aac.02325-18

Bullard-Feibelman KM, Fuller BP, Geiss BJ (2016) A sensitive and robust high-throughput screening assay for inhibitors of the Chikungunya virus nsP1 capping enzyme. PLoS ONE 11 (7):e0158923. https://doi.org/10.1371/journal.pone.0158923

Cassadou S, Boucau S, Petit-Sinturel M, Huc P, Leparc-Goffart I, Ledrans M (2014) Emergence of Chikungunya fever on the French side of Saint Martin island, October–December 2013. Euro Surveill 19(13). https://doi.org/10.2807/1560-7917.es2014.19.13.20752

Chang SY, Bae SJ, Lee MY, Baek SH, Chang S, Kim SH (2011) Chemical affinity matrix-based identification of prohibitin as a binding protein to anti-resorptive sulfonyl amidine compounds. Bioorg Med Chem Lett 21(2):727–729. https://doi.org/10.1016/j.bmcl.2010.11.123

Charrel RN, de Lamballerie X, Raoult D (2007) Chikungunya outbreaks–the globalization of vectorborne diseases. N Engl J Med 356(8):769–771. https://doi.org/10.1056/NEJMp078013

Cho B, Jeon BY, Kim J, Noh J, Kim J, Park M et al (2008) Expression and evaluation of Chikungunya virus E1 and E2 envelope proteins for serodiagnosis of Chikungunya virus infection. Yonsei Med J 49(5):828–835. https://doi.org/10.3349/ymj.2008.49.5.828

Choi HK, Tong L, Minor W, Dumas P, Boege U, Rossmann MG et al (1991) Structure of Sindbis virus core protein reveals a chymotrypsin-like serine proteinase and the organization of the virion. Nature 354(6348):37–43. https://doi.org/10.1038/354037a0

Chopra A, Anuradha V, Ghorpade R, Saluja M (2012) Acute Chikungunya and persistent musculoskeletal pain following the 2006 Indian epidemic: a 2-year prospective rural community study. Epidemiol Infect 140(5):842–850. https://doi.org/10.1017/S0950268811001300

Chung DH, Golden JE, Adcock RS, Schroeder CE, Chu YK, Sotsky JB et al (2016) Discovery of a broad-spectrum antiviral compound that inhibits pyrimidine biosynthesis and establishes a type 1 interferon-independent antiviral state. Antimicrob Agents Chemother 60(8):4552–4562. https://doi.org/10.1128/AAC.00282-16

Colpitts CC, Schang LM (2014) A small molecule inhibits virion attachment to heparan sulfate- or sialic acid-containing glycans. J Virol 88(14):7806–7817. https://doi.org/10.1128/JVI.00896-14

Constantinescu SN, Cernescu CD, Popescu LM (1991) Effects of protein kinase C inhibitors on viral entry and infectivity. FEBS Lett 292(1–2):31–33. https://doi.org/10.1016/0014-5793(91)80826-o

Crutcher WA, Moschella SL (1975) Double-blind controlled crossover high-dose study of Azaribine in psoriasis. Br J Dermatol 92(2):199–205. https://doi.org/10.1111/j.1365-2133.1975.tb03059.x

Cruz DJ, Bonotto RM, Gomes RG, da Silva CT, Taniguchi JB, No JH et al (2013) Identification of novel compounds inhibiting Chikungunya virus-induced cell death by high throughput screening of a kinase inhibitor library. PLoS Negl Trop Dis 7(10):e2471. https://doi.org/10.1371/journal.pntd.0002471

Das I, Basantray I, Mamidi P, Nayak TK, Pratheek BM, Chattopadhyay S, Chattopadhyay S (2014a) Heat shock protein 90 positively regulates Chikungunya virus replication by stabilizing viral non-structural protein nsP2 during infection. PLoS One. 9(6):e100531. https://doi.org/10.1371/journal.pone.0100531

Das PK, Merits A, Lulla A (2014b) Functional cross-talk between distant domains of Chikungunya virus non-structural protein 2 is decisive for its RNA-modulating activity. J Biol Chem 289(9):5635–5653. https://doi.org/10.1074/jbc.M113.503433

Das PK, Puusepp L, Varghese FS, Utt A, Ahola T, Kananovich DG et al (2016) Design and validation of novel Chikungunya virus protease inhibitors. Antimicrob Agents Chemother 60 (12):7382–7395. https://doi.org/10.1128/AAC.01421-16

De Clercq E (2007) The design of drugs for HIV and HCV. Nat Rev Drug Discov 6(12):1001–1018. https://doi.org/10.1038/nrd2424

de Groot RJ, Hardy WR, Shirako Y, Strauss JH (1990) Cleavage-site preferences of Sindbis virus polyproteins containing the non-structural proteinase. Evidence for temporal regulation of polyprotein processing in vivo. EMBO J 9(8):2631–2638

Delang L, Li C, Tas A, Querat G, Albulescu IC, De Burghgraeve T et al (2016) The viral capping enzyme nsP1: a novel target for the inhibition of Chikungunya virus infection. Sci Rep 6:31819. https://doi.org/10.1038/srep31819

Delang L, Segura Guerrero N, Tas A, Querat G, Pastorino B, Froeyen M et al (2014) Mutations in the Chikungunya virus non-structural proteins cause resistance to favipiravir (T-705), a broad-spectrum antiviral. J Antimicrob Chemother 69(10):2770–2784. https://doi.org/10.1093/jac/dku209

Delogu I, Pastorino B, Baronti C, Nougairede A, Bonnet E, de Lamballerie X (2011) In vitro antiviral activity of arbidol against Chikungunya virus and characteristics of a selected resistant mutant. Antiviral Res 90(3):99–107. https://doi.org/10.1016/j.antiviral.2011.03.182

Deneau DG, Farber EM (1975) The treatment of psoriasis with azaribine. Dermatologica 151 (3):158–163

Dhindwal S, Kesari P, Singh H, Kumar P, Tomar S (2017) Conformer and pharmacophore based identification of peptidomimetic inhibitors of Chikungunya virus nsP2 protease. J Biomol Struct Dyn 35(16):3522–3539. https://doi.org/10.1080/07391102.2016.1261046

Di Mola A, Peduto A, La Gatta A, Delang L, Pastorino B, Neyts J et al (2014) Structure-activity relationship study of arbidol derivatives as inhibitors of Chikungunya virus replication. Bioorg Med Chem 22(21):6014–6025. https://doi.org/10.1016/j.bmc.2014.09.013

Dryga SA, Dryga OA, Schlesinger S (1997) Identification of mutations in a sindbis virus variant able to establish persistent infection in BHK cells: the importance of a mutation in the nsP2 gene. Virology 228(1):74–83. https://doi.org/10.1006/viro.1996.8364

Dubin DT, Stollar V (1975) Methylation of sindbis virus "26S" messenger RNA. Biochem Biophys Res Commun 66(4):1373–1379. https://doi.org/10.1016/0006-291x(75)90511-2

Dubin DT, Stollar V, Hsuchen CC, Timko K, Guild GM (1977) Sindbis virus messenger RNA: the 5′-termini and methylated residues of 26 and 42 S RNA. Virology 77(2):457–470. https://doi.org/10.1016/0042-6822(77)90471-8

Ehteshami M, Tao S, Zandi K, Hsiao HM, Jiang Y, Hammond E et al (2017) Characterization of β-d-N(4)-Hydroxycytidine as a novel inhibitor of Chikungunya virus. Antimicrob Agents Chemother 61(4). https://doi.org/10.1128/aac.02395-16

Ekins S, Mestres J, Testa B (2007a) In silico pharmacology for drug discovery: applications to targets and beyond. Br J Pharmacol 152(1):21–37. https://doi.org/10.1038/sj.bjp.0707306

Ekins S, Mestres J, Testa B (2007b) In silico pharmacology for drug discovery: methods for virtual ligand screening and profiling. Br J Pharmacol 152(1):9–20. https://doi.org/10.1038/sj.bjp.0707305

Feibelman KM, Fuller BP, Li L, LaBarbera DV, Geiss BJ (2018) Identification of small molecule inhibitors of the Chikungunya virus nsP1 RNA capping enzyme. Antiviral Res 154:124–131. https://doi.org/10.1016/j.antiviral.2018.03.013

Ferreira AC, Reis PA, de Freitas CS, Sacramento CQ, Villas Boas Hoelz L, Bastos MM et al (2019) Beyond members of the Flaviviridae family, sofosbuvir also inhibits Chikungunya virus replication. Antimicrob Agents Chemother. 63(2). https://doi.org/10.1128/aac.01389-18

Fresno M, Jimenez A, Vazquez D (1977) Inhibition of translation in eukaryotic systems by harringtonine. Eur J Biochem/FEBS 72(2):323–330. https://doi.org/10.1111/j.1432-1033.1977.tb11256.x

Frolov I, Agapov E, Hoffman TA Jr, Pragai BM, Lippa M, Schlesinger S et al (1999) Selection of RNA replicons capable of persistent noncytopathic replication in mammalian cells. J Virol 73 (5):3854–3865

Frolov I, Garmashova N, Atasheva S, Frolova EI (2009) Random insertion mutagenesis of sindbis virus nonstructural protein 2 and selection of variants incapable of downregulating cellular transcription. J Virol 83(18):9031–9044. https://doi.org/10.1128/JVI.00850-09

Frolova EI, Gorchakov R, Pereboeva L, Atasheva S, Frolov I (2010) Functional Sindbis virus replicative complexes are formed at the plasma membrane. J Virol 84(22):11679–11695. https://doi.org/10.1128/JVI.01441-10

Fros JJ, Domeradzka NE, Baggen J, Geertsema C, Flipse J, Vlak JM et al (2012) Chikungunya virus nsP3 blocks stress granule assembly by recruitment of G3BP into cytoplasmic foci. J Virol 86(19):10873–10879. https://doi.org/10.1128/JVI.01506-12

Fros JJ, Geertsema C, Zouache K, Baggen J, Domeradzka N, van Leeuwen DM et al (2015) Mosquito Rasputin interacts with Chikungunya virus nsP3 and determines the infection rate in *Aedes albopictus*. Parasit Vectors 8:464. https://doi.org/10.1186/s13071-015-1070-4

Furuta Y, Gowen BB, Takahashi K, Shiraki K, Smee DF, Barnard DL (2013) Favipiravir (T-705), a novel viral RNA polymerase inhibitor. Antiviral Res 100(2):446–454. https://doi.org/10.1016/j.antiviral.2013.09.015

Furuta Y, Komeno T, Nakamura T (2017) Favipiravir (T-705), a broad spectrum inhibitor of viral RNA polymerase. Proc Jpn Acad Ser B Phys Biol Sci 93(7):449–463. https://doi.org/10.2183/pjab.93.027

Furuta Y, Takahashi K, Kuno-Maekawa M, Sangawa H, Uehara S, Kozaki K et al (2005) Mechanism of action of T-705 against influenza virus. Antimicrob Agents Chemother 49 (3):981–986. https://doi.org/10.1128/AAC.49.3.981-986.2005

Gao Y, Goonawardane N, Ward J, Tuplin A, Harris M (2019) Multiple roles of the non-structural protein 3 (nsP3) alphavirus unique domain (AUD) during Chikungunya virus genome replication and transcription. PLoS Pathog 15(1):e1007239. https://doi.org/10.1371/journal.ppat.1007239

Garoff H, Sjoberg M, Cheng RH (2004) Budding of alphaviruses. Virus Res 106(2):103–116. https://doi.org/10.1016/j.virusres.2004.08.008

Gerardin P, Guernier V, Perrau J, Fianu A, Le Roux K, Grivard P et al (2008) Estimating Chikungunya prevalence in La Reunion Island outbreak by serosurveys: two methods for two critical times of the epidemic. BMC Infect Dis 8:99. https://doi.org/10.1186/1471-2334-8-99

Gigante A, Canela MD, Delang L, Priego EM, Camarasa MJ, Querat G et al (2014) Identification of [1,2,3]triazolo[4,5-d]pyrimidin-7(6H)-ones as novel inhibitors of Chikungunya virus replication. J Med Chem 57(10):4000–4008. https://doi.org/10.1021/jm401844c

Gigante A, Gomez-SanJuan A, Delang L, Li C, Bueno O, Gamo AM et al (2017) Antiviral activity of [1,2,3]triazolo[4,5-d]pyrimidin-7(6H)-ones against Chikungunya virus targeting the viral capping nsP1. Antiviral Res 144:216–222. https://doi.org/10.1016/j.antiviral.2017.06.003

Gopakumar H, Ramachandran S (2012) Congenital Chikungunya. J Clin Neonatol 1(3):155–156. https://doi.org/10.4103/2249-4847.101704

Gorchakov R, Frolova E, Frolov I (2005) Inhibition of transcription and translation in Sindbis virus-infected cells. J Virol 79(15):9397–9409

Gupta DK, Kaur P, Leong ST, Tan LT, Prinsep MR, Chu JJ (2014) Anti-Chikungunya viral activities of aplysiatoxin-related compounds from the marine cyanobacterium *Trichodesmium erythraeum*. Mar Drugs 12(1):115–127. https://doi.org/10.3390/md12010115

Hallenberger S, Bosch V, Angliker H, Shaw E, Klenk HD, Garten W (1992) Inhibition of furin-mediated cleavage activation of HIV-1 glycoprotein gp160. Nature 360(6402):358–361. https://doi.org/10.1038/360358a0

Heidner HW, Knott TA, Johnston RE (1996) Differential processing of sindbis virus glycoprotein PE2 in cultured vertebrate and arthropod cells. J Virol 70(3):2069–2073

Heidner HW, McKnight KL, Davis NL, Johnston RE (1994) Lethality of PE2 incorporation into Sindbis virus can be suppressed by second-site mutations in E3 and E2. J Virol 68(4):2683–2692

Ho YJ, Wang YM, Lu JW, Wu TY, Lin LI, Kuo SC et al (2015) Suramin inhibits Chikungunya virus entry and transmission. PLoS ONE 10(7):e0133511. https://doi.org/10.1371/journal.pone.0133511

Hoffmann HH, Kunz A, Simon VA, Palese P, Shaw ML (2011) Broad-spectrum antiviral that interferes with de novo pyrimidine biosynthesis. Proc Natl Acad Sci USA 108(14):5777–5782. https://doi.org/10.1073/pnas.1101143108

Hoornweg TE, van Duijl-Richter MKS, Ayala Nunez NV, Albulescu IC, van Hemert MJ, Smit JM (2016) Dynamics of Chikungunya virus cell entry unraveled by single-virus tracking in living cells. J Virol 90(9):4745–4756. https://doi.org/10.1128/JVI.03184-15

Jacobs SC, Taylor A, Herrero LJ, Mahalingam S, Fazakerley JK 2017 (2017) Mutation of a conserved nuclear export sequence in Chikungunya virus capsid protein disrupts host cell nuclear import. Viruses 9(10). https://doi.org/10.3390/v9100306

Jadav SS, Sinha BN, Hilgenfeld R, Jayaprakash V (2017) Computer-aided structure based drug design approaches for the discovery of new anti-CHIKV agents. Curr Comput Aided Drug Des 13(4):346–361. https://doi.org/10.2174/1573409913666170309145308

Jadav SS, Sinha BN, Hilgenfeld R, Pastorino B, de Lamballerie X, Jayaprakash V (2015) Thiazolidone derivatives as inhibitors of Chikungunya virus. Eur J Med Chem 89:172–178. https://doi.org/10.1016/j.ejmech.2014.10.042

Jhaveri K, Modi S (2015) Ganetespib: research and clinical development. Onco Targets Ther 8:1849–1858. https://doi.org/10.2147/OTT.S65804

Jones PH, Maric M, Madison MN, Maury W, Roller RJ, Okeoma CM (2013) BST-2/tetherin-mediated restriction of Chikungunya (CHIKV) VLP budding is counteracted by CHIKV non-structural protein 1 (nsP1). Virology 438(1):37–49. https://doi.org/10.1016/j.virol.2013.01.010

Jose J, Przybyla L, Edwards TJ, Perera R, Burgner JW 2nd, Kuhn RJ (2012) Interactions of the cytoplasmic domain of Sindbis virus E2 with nucleocapsid cores promote alphavirus budding. J Virol 86(5):2585–2599. https://doi.org/10.1128/JVI.05860-11

Jose J, Snyder JE, Kuhn RJ (2009) A structural and functional perspective of alphavirus replication and assembly. Future Microbiol 4(7):837–856. https://doi.org/10.2217/fmb.09.59

Kaihatsu K, Yamabe M, Ebara Y (2018) Antiviral mechanism of action of epigallocatechin-3-*O*-gallate and its fatty acid esters. Molecules 23(10). https://doi.org/10.3390/molecules23102475

Karlas A, Berre S, Couderc T, Varjak M, Braun P, Meyer M et al (2016) A human genome-wide loss-of-function screen identifies effective Chikungunya antiviral drugs. Nat Commun 7:11320. https://doi.org/10.1038/ncomms11320

Karpe YA, Aher PP, Lole KS (2011) NTPase and 5′-RNA triphosphatase activities of Chikungunya virus nsP2 protein. PLoS ONE 6(7):e22336. https://doi.org/10.1371/journal.pone.0022336

Kaur P, Thiruchelvan M, Lee RC, Chen H, Chen KC, Ng ML et al (2013) Inhibition of Chikungunya virus replication by harringtonine, a novel antiviral that suppresses viral protein expression. Antimicrob Agents Chemother 57(1):155–167. https://doi.org/10.1128/AAC.01467-12

Keating JA, Striker R (2012) Phosphorylation events during viral infections provide potential therapeutic targets. Rev Med Virol 22(3):166–181. https://doi.org/10.1002/rmv.722

Khan M, Dhanwani R, Patro IK, Rao PV, Parida MM (2011) Cellular IMPDH enzyme activity is a potential target for the inhibition of Chikungunya virus replication and virus induced apoptosis in cultured mammalian cells. Antiviral Res 89(1):1–8. https://doi.org/10.1016/j.antiviral.2010.10.009

Khan M, Santhosh SR, Tiwari M, Lakshmana Rao PV, Parida M (2010) Assessment of in vitro prophylactic and therapeutic efficacy of chloroquine against Chikungunya virus in vero cells. J Med Virol 82(5):817–824. https://doi.org/10.1002/jmv.21663

Kielian M, Chanel-Vos C, Liao M (2010) Alphavirus entry and membrane fusion. Viruses 2 (4):796–825. https://doi.org/10.3390/v2040796

Kim HY, Patkar C, Warrier R, Kuhn R, Cushman M (2005) Design, synthesis, and evaluation of dioxane-based antiviral agents targeted against the Sindbis virus capsid protein. Bioorg Med Chem Lett 15(13):3207–3211. https://doi.org/10.1016/j.bmcl.2005.05.013

Kim KH, Rumenapf T, Strauss EG, Strauss JH (2004) Regulation of Semliki Forest virus RNA replication: a model for the control of alphavirus pathogenesis in invertebrate hosts. Virology 323(1):153–163. https://doi.org/10.1016/j.virol.2004.03.009

Klimstra WB, Heidner HW, Johnston RE (1999) The furin protease cleavage recognition sequence of sindbis virus PE2 can mediate virion attachment to cell surface heparan sulfate. J Virol 73 (8):6299–6306

Langsjoen RM, Auguste AJ, Rossi SL, Roundy CM, Penate HN, Kastis M et al (2017) Host oxidative folding pathways offer novel anti-Chikungunya virus drug targets with broad spectrum potential. Antiviral Res 143:246–251. https://doi.org/10.1016/j.antiviral.2017.04.014

Lark T, Keck F, Narayanan A (2017) Interactions of alphavirus nsP3 protein with host proteins. Front Microbiol 8:2652. https://doi.org/10.3389/fmicb.2017.02652

Lebrun G, Chadda K, Reboux AH, Martinet O, Gauzere BA (2009) Guillain-Barre syndrome after Chikungunya infection. Emerg Infect Dis 15(3):495–496. https://doi.org/10.3201/eid1503.071482

Lee CHR, Mohamed Hussain K, Chu JJH (2019) Macropinocytosis dependent entry of Chikungunya virus into human muscle cells. PLoS Negl Trop Dis 13(8):e0007610. https://doi.org/10.1371/journal.pntd.0007610

Leyssen P, De Clercq E, Neyts J (2006) The anti-yellow fever virus activity of ribavirin is independent of error-prone replication. Mol Pharmacol 69(4):1461–1467. https://doi.org/10.1124/mol.105.020057

Li L, Jose J, Xiang Y, Kuhn RJ, Rossmann MG (2010) Structural changes of envelope proteins during alphavirus fusion. Nature 468(7324):705–708. https://doi.org/10.1038/nature09546

Linger BR, Kunovska L, Kuhn RJ, Golden BL (2004) Sindbis virus nucleocapsid assembly: RNA folding promotes capsid protein dimerization. RNA 10(1):128–138. https://doi.org/10.1261/rna.5127104

Lu JW, Hsieh PS, Lin CC, Hu MK, Huang SM, Wang YM et al (2017) Synergistic effects of combination treatment using EGCG and suramin against the Chikungunya virus. Biochem Biophys Res Commun 491(3):595–602. https://doi.org/10.1016/j.bbrc.2017.07.157

Lucas-Hourani M, Dauzonne D, Jorda P, Cousin G, Lupan A, Helynck O et al (2013a) Inhibition of pyrimidine biosynthesis pathway suppresses viral growth through innate immunity. PLoS Pathog 9(10):e1003678. https://doi.org/10.1371/journal.ppat.1003678

Lucas-Hourani M, Lupan A, Despres P, Thoret S, Pamlard O, Dubois J et al (2013b) A phenotypic assay to identify Chikungunya virus inhibitors targeting the nonstructural protein nsP2. J Biomol Screen 18(2):172–179. https://doi.org/10.1177/1087057112460091

Lyra PP, Campos GS, Bandeira ID, Sardi SI, Costa LF, Santos FR et al (2016) Congenital Chikungunya virus infection after an outbreak in Salvador, Bahia, Brazil. AJP Rep 6(3):e299–e300. https://doi.org/10.1055/s-0036-1587323

Machkovech HM, Bloom JD, Subramaniam AR (2019) Comprehensive profiling of translation initiation in influenza virus infected cells. PLoS Pathog 15(1):e1007518. https://doi.org/10.1371/journal.ppat.1007518

Malet H, Coutard B, Jamal S, Dutartre H, Papageorgiou N, Neuvonen M et al (2009) The crystal structures of Chikungunya and venezuelan equine encephalitis virus nsP3 macro domains define a conserved adenosine binding pocket. J Virol 83(13):6534–6545. https://doi.org/10.1128/JVI.00189-09

Manns MP, von Hahn T (2013) Novel therapies for hepatitis C—one pill fits all? Nat Rev Drug Discov 12(8):595–610. https://doi.org/10.1038/nrd4050

McPherson RL, Abraham R, Sreekumar E, Ong SE, Cheng SJ, Baxter VK et al (2017) ADP-ribosylhydrolase activity of Chikungunya virus macrodomain is critical for virus replication and virulence. Proc Natl Acad Sci USA 114(7):1666–1671. https://doi.org/10.1073/pnas.1621485114

Melancon P, Garoff H (1987) Processing of the Semliki Forest virus structural polyprotein: role of the capsid protease. J Virol 61(5):1301–1309

Metz SW, Pijlman GP (2016a) Function of Chikungunya virus structural proteins. In: Okeoma CM (ed) Chikungunya virus advances in biology, pathogenesis, and treatment, pp 63–74. Switzerland: Springer, Cham

Metz SW, Pijlman GP (2016b) Production of Chikungunya virus-like particles and subunit vaccines in insect cells. Methods Mol Biol 1426:297–309. https://doi.org/10.1007/978-1-4939-3618-2_27

Mishra P, Kumar A, Mamidi P, Kumar S, Basantray I, Saswat T et al (2016) Inhibition of Chikungunya virus replication by 1-[(2-Methylbenzimidazol-1-yl) Methyl]-2-Oxo-Indolin-3-ylidene] Amino] Thiourea(MBZM-N-IBT). Sci Rep 6:20122. https://doi.org/10.1038/srep20122

Moller-Tank S, Kondratowicz AS, Davey RA, Rennert PD, Maury W (2013) Role of the phosphatidylserine receptor TIM-1 in enveloped-virus entry. J Virol 87(15):8327–8341. https://doi.org/10.1128/JVI.01025-13

Mukhopadhyay S, Zhang W, Gabler S, Chipman PR, Strauss EG, Strauss JH et al (2006) Mapping the structure and function of the E1 and E2 glycoproteins in alphaviruses. Structure 14(1):63–73. https://doi.org/10.1016/j.str.2005.07.025

Nayak TK, Mamidi P, Kumar A, Singh LP, Sahoo SS, Chattopadhyay S et al (2017) Regulation of viral replication, apoptosis and pro-inflammatory responses by 17-AAG during Chikungunya virus infection in macrophages. Viruses 9(1). https://doi.org/10.3390/v9010003

Nguyen PT, Yu H, Keller PA (2014) Discovery of in silico hits targeting the nsP3 macro domain of Chikungunya virus. J Mol Model 20(5):2216. https://doi.org/10.1007/s00894-014-2216-6

Nguyen PT, Yu H, Keller PA (2015) Identification of Chikungunya virus nsP2 protease inhibitors using structure-base approaches. J Mol Graph Model 57:1–8. https://doi.org/10.1016/j.jmgm.2015.01.001

Ooi YS, Stiles KM, Liu CY, Taylor GM, Kielian M (2013) Genome-wide RNAi screen identifies novel host proteins required for alphavirus entry. PLoS Pathog 9(12):e1003835. https://doi.org/10.1371/journal.ppat.1003835

Ortiz-Riano E, Ngo N, Devito S, Eggink D, Munger J, Shaw ML et al (2014) Inhibition of arenavirus by A3, a pyrimidine biosynthesis inhibitor. J Virol 88(2):878–889. https://doi.org/10.1128/JVI.02275-13

Ozden S, Lucas-Hourani M, Ceccaldi PE, Basak A, Valentine M, Benjannet S et al (2008) Inhibition of Chikungunya virus infection in cultured human muscle cells by furin inhibitors: impairment of the maturation of the E2 surface glycoprotein. J Biol Chem 283(32):21899–21908. doi: M802444200 [pii] https://doi.org/10.1074/jbc.m802444200

Paeshuyse J, Dallmeier K, Neyts J (2011) Ribavirin for the treatment of chronic hepatitis C virus infection: a review of the proposed mechanisms of action. Curr Opin Virol 1(6):590–598. https://doi.org/10.1016/j.coviro.2011.10.030

Panas MD, Ahola T, McInerney GM (2014) The C-terminal repeat domains of nsP3 from the old world alphaviruses bind directly to G3BP. J Virol 88(10):5888–5893. https://doi.org/10.1128/JVI.00439-14

Panas MD, Varjak M, Lulla A, Eng KE, Merits A, Karlsson Hedestam GB et al (2012) Sequestration of G3BP coupled with efficient translation inhibits stress granules in Semliki Forest virus infection. Mol Biol Cell 23(24):4701–4712. https://doi.org/10.1091/mbc.E12-08-0619

Pawlotsky JM (2014) New hepatitis C virus (HCV) drugs and the hope for a cure: concepts in anti-HCV drug development. Semin Liver Dis 34(1):22–29. https://doi.org/10.1055/s-0034-1371007

Pietila MK, Hellstrom K, Ahola T (2017) Alphavirus polymerase and RNA replication. Virus Res 234:44–57. https://doi.org/10.1016/j.virusres.2017.01.007

Pohjala L, Utt A, Varjak M, Lulla A, Merits A, Ahola T et al (2011) Inhibitors of alphavirus entry and replication identified with a stable Chikungunya replicon cell line and virus-based assays. PLoS ONE 6(12):e28923. https://doi.org/10.1371/journal.pone.0028923

Pokorna J, Machala L, Rezacova P, Konvalinka J (2009) Current and novel inhibitors of HIV protease. Viruses 1(3):1209–1239. https://doi.org/10.3390/v1031209

Rada B, Dragun M (1977) Antiviral action and selectivity of 6-azauridine. Ann NY Acad Sci 284:410–417. https://doi.org/10.1111/j.1749-6632.1977.tb21977.x

Ramakrishnan C, Kutumbarao NHV, Suhitha S, Velmurugan D (2017) Structure-function relationship of Chikungunya nsP2 protease: a comparative study with papain. Chem Biol Drug Des 89(5):772–782. https://doi.org/10.1111/cbdd.12901

Rathore AP, Haystead T, Das PK, Merits A, Ng ML, Vasudevan SG (2014) Chikungunya virus nsP3 & nsP4 interacts with HSP-90 to promote virus replication: HSP-90 inhibitors reduce CHIKV infection and inflammation in vivo. Antiviral Res 103:7–16. https://doi.org/10.1016/j.antiviral.2013.12.010

Rathore AP, Ng ML, Vasudevan SG (2013) Differential unfolded protein response during Chikungunya and Sindbis virus infection: CHIKV nsP4 suppresses eIF2alpha phosphorylation. Virol J 10:36. https://doi.org/10.1186/1743-422X-10-36

Raveh A, Delekta PC, Dobry CJ, Peng W, Schultz PJ, Blakely PK et al (2013) Discovery of potent broad spectrum antivirals derived from marine actinobacteria. PLoS ONE 8(12):e82318. https://doi.org/10.1371/journal.pone.0082318

Reid SP, Tritsch SR, Kota K, Chiang CY, Dong L, Kenny T et al (2015) Sphingosine kinase 2 is a Chikungunya virus host factor co-localized with the viral replication complex. Emerg Microbes Infect 4(10):e61. https://doi.org/10.1038/emi.2015.61

Rezza G, Nicoletti L, Angelini R, Romi R, Finarelli AC, Panning M et al (2007) Infection with Chikungunya virus in Italy: an outbreak in a temperate region. Lancet 370(9602):1840–1846

Ribeiro Morais G, Vicente Miranda H, Santos IC, Santos I, Outeiro TF, Paulo A (2011) Synthesis and in vitro evaluation of fluorinated styryl benzazoles as amyloid-probes. Bioorg Med Chem 19(24):7698–7710. https://doi.org/10.1016/j.bmc.2011.09.065

Roques P, Thiberville SD, Dupuis-Maguiraga L, Lum FM, Labadie K, Martinon F et al (2018) Paradoxical effect of chloroquine treatment in enhancing Chikungunya virus infection. Viruses 10(5). https://doi.org/10.3390/v10050268

Rothan HA, Bahrani H, Mohamed Z, Teoh TC, Shankar EM, Rahman NA et al (2015) A combination of doxycycline and ribavirin alleviated Chikungunya infection. PLoS ONE 10(5): e0126360. https://doi.org/10.1371/journal.pone.0126360

Rubach JK, Wasik BR, Rupp JC, Kuhn RJ, Hardy RW, Smith JL (2009) Characterization of purified sindbis virus nsP4 RNA-dependent RNA polymerase activity in vitro. Virology 384 (1):201–208. https://doi.org/10.1016/j.virol.2008.10.030

Rupp JC, Sokoloski KJ, Gebhart NN, Hardy RW (2015) Alphavirus RNA synthesis and non-structural protein functions. J Gen Virol 96(9):2483–2500. https://doi.org/10.1099/jgv.0.000249

Sahoo B, Chowdary TK (2019) Conformational changes in Chikungunya virus E2 protein upon heparan sulfate receptor binding explain mechanism of E2-E1 dissociation during viral entry. Biosci Rep 39(6). https://doi.org/10.1042/bsr20191077

Salonen A, Vasiljeva L, Merits A, Magden J, Jokitalo E, Kaariainen L (2003) Properly folded nonstructural polyprotein directs the Semliki Forest virus replication complex to the endosomal compartment. J Virol 77(3):1691–1702. https://doi.org/10.1128/jvi.77.3.1691-1702.2003

Savarino A, Boelaert JR, Cassone A, Majori G, Cauda R (2003) Effects of chloroquine on viral infections: an old drug against today's diseases? Lancet Infect Dis 3(11):722–727

Scholte FE, Tas A, Martina BE, Cordioli P, Narayanan K, Makino S et al (2013) Characterization of synthetic Chikungunya viruses based on the consensus sequence of recent E1-226V isolates. PLoS ONE 8(8):e71047. https://doi.org/10.1371/journal.pone.0071047

Schuffenecker I, Iteman I, Michault A, Murri S, Frangeul L, Vaney MC et al (2006) Genome microevolution of Chikungunya viruses causing the Indian Ocean outbreak. PLoS Med 3(7): e263. https://doi.org/10.1371/journal.pmed.0030263

Seyedi SS, Shukri M, Hassandarvish P, Oo A, Shankar EM, Abubakar S et al (2016) Computational approach towards exploring potential anti-Chikungunya activity of selected flavonoids. Scientific reports 6:24027. https://doi.org/10.1038/srep24027

Sharma R, Fatma B, Saha A, Bajpai S, Sistla S, Dash PK et al (2016) Inhibition of Chikungunya virus by picolinate that targets viral capsid protein. Virology 498:265–276. https://doi.org/10.1016/j.virol.2016.08.029

Sharma R, Kesari P, Kumar P, Tomar S (2018) Structure-function insights into Chikungunya virus capsid protein: small molecules targeting capsid hydrophobic pocket. Virology 515:223–234. https://doi.org/10.1016/j.virol.2017.12.020

Shirako Y, Strauss JH (1994) Regulation of sindbis virus RNA replication: uncleaved P123 and nsP4 function in minus-strand RNA synthesis, whereas cleaved products from P123 are required for efficient plus-strand RNA synthesis. J Virol 68(3):1874–1885

Silva LA, Khomandiak S, Ashbrook AW, Weller R, Heise MT, Morrison TE et al (2014) A single-amino-acid polymorphism in Chikungunya virus E2 glycoprotein influences glycosaminoglycan utilization. J Virol 88(5):2385–2397. https://doi.org/10.1128/JVI.03116-13

Simmons DT, Strauss JH (1972) Replication of Sindbis virus. I. Relative size and genetic content of 26 s and 49 s RNA. J Mol Biol 71(3):599–613

Simon F, Javelle E, Cabie A, Bouquillard E, Troisgros O, Gentile G et al (2015) French guidelines for the management of Chikungunya (acute and persistent presentations). November 2014. Med Mal Infect 45(7):243–263. https://doi.org/10.1016/j.medmal.2015.05.007

Singh H, Mudgal R, Narwal M, Kaur R, Singh VA, Malik A et al (2018) Chikungunya virus inhibition by peptidomimetic inhibitors targeting virus-specific cysteine protease. Biochimie 149:51–61. https://doi.org/10.1016/j.biochi.2018.04.004

Sissoko D, Malvy D, Ezzedine K, Renault P, Moscetti F, Ledrans M et al (2009) Post-epidemic Chikungunya disease on reunion island: course of rheumatic manifestations and associated factors over a 15-month period. PLoS Negl Trop Dis 3(3):e389. https://doi.org/10.1371/journal.pntd.0000389

Smee DF, Hurst BL, Day CW (2012) D282, a non-nucleoside inhibitor of influenza virus infection that interferes with de novo pyrimidine biosynthesis. Antivir Chem Chemother 22(6):263–272. https://doi.org/10.3851/IMP2105

Sokoloski KJ, Nease LM, May NA, Gebhart NN, Jones CE, Morrison TE et al (2017) Identification of interactions between sindbis virus capsid protein and cytoplasmic vRNA as novel virulence determinants. PLoS Pathog 13(6):e1006473. https://doi.org/10.1371/journal.ppat.1006473

Solignat M, Gay B, Higgs S, Briant L, Devaux C (2009) Replication cycle of Chikungunya: a re-emerging arbovirus. Virology 393(2):183–197. https://doi.org/10.1016/j.virol.2009.07.024

Spuul P, Balistreri G, Kaariainen L, Ahola T (2010) Phosphatidylinositol 3-kinase-, actin-, and microtubule-dependent transport of Semliki Forest virus replication complexes from the plasma membrane to modified lysosomes. J Virol 84(15):7543–7557. https://doi.org/10.1128/JVI.00477-10

Spuul P, Salonen A, Merits A, Jokitalo E, Kaariainen L, Ahola T (2007) Role of the amphipathic peptide of Semliki Forest virus replicase protein nsP1 in membrane association and virus replication. J Virol 81(2):872–883. https://doi.org/10.1128/JVI.01785-06

Stangherlin LM, de Paula FN, Icimoto MY, Ruiz LGP, Nogueira ML, Braz ASK et al (2017) Positively selected sites at HCMV gB furin processing region and their effects in cleavage efficiency. Frontiers in microbiology 8:934. https://doi.org/10.3389/fmicb.2017.00934

Staveness D, Abdelnabi R, Near KE, Nakagawa Y, Neyts J, Delang L et al (2016) Inhibition of Chikungunya virus-induced cell death by salicylate-derived bryostatin analogues provides additional evidence for a PKC-independent pathway. J Nat Prod 79(4):680–684. https://doi.org/10.1021/acs.jnatprod.5b01017

Strauss EG, Rice CM, Strauss JH (1984) Complete nucleotide sequence of the genomic RNA of sindbis virus. Virology 133(1):92–110. https://doi.org/10.1016/0042-6822(84)90428-8

Strauss JH, Strauss EG (1994) The alphaviruses: gene expression, replication, and evolution. Microbiol Rev 58(3):491–562

Subudhi BB, Chattopadhyay S, Mishra P, Kumar A (2018) Current strategies for inhibition of Chikungunya infection. Viruses 10(5). https://doi.org/10.3390/v10050235

Tamm K, Merits A, Sarand I (2008) Mutations in the nuclear localization signal of nsP2 influencing RNA synthesis, protein expression and cytotoxicity of Semliki Forest virus. J Gen Virol 89(Pt 3):676–686. https://doi.org/10.1099/vir.0.83320-0

Taylor A, Liu X, Zaid A, Goh LY, Hobson-Peters J, Hall RA et al (2017) Mutation of the *N*-terminal region of Chikungunya virus capsid protein: implications for vaccine design. mBio 8 (1). https://doi.org/10.1128/mbio.01970-16

Thaa B, Biasiotto R, Eng K, Neuvonen M, Gotte B, Rheinemann L et al (2015) Differential phosphatidylinositol-3-kinase-Akt-mTOR activation by Semliki Forest and Chikungunya viruses is dependent on nsP3 and connected to replication complex internalization. J Virol 89 (22):11420–11437. https://doi.org/10.1128/JVI.01579-15

Thomas S, Rai J, John L, Gunther S, Drosten C, Putzer BM et al (2010) Functional dissection of the alphavirus capsid protease: sequence requirements for activity. Virol J 7:327. https://doi.org/10.1186/1743-422X-7-327

Thomas S, Rai J, John L, Schaefer S, Putzer BM, Herchenroder O (2013) Chikungunya virus capsid protein contains nuclear import and export signals. Virol J 10:269. https://doi.org/10.1186/1743-422X-10-269

Tomar S, Hardy RW, Smith JL, Kuhn RJ (2006) Catalytic core of alphavirus nonstructural protein nsP4 possesses terminal adenylyltransferase activity. J Virol 80(20):9962–9969. https://doi.org/10.1128/JVI.01067-06

Tsetsarkin KA, Vanlandingham DL, McGee CE, Higgs S (2007) A single mutation in Chikungunya virus affects vector specificity and epidemic potential. PLoS Pathog 3(12):e201

Turner TL, Kopp BT, Paul G, Landgrave LC, Hayes D Jr, Thompson R (2014) Respiratory syncytial virus: current and emerging treatment options. Clinicoecon Outcomes Res 6:217–225. https://doi.org/10.2147/CEOR.S60710

Uchime O, Fields W, Kielian M (2013) The role of E3 in pH protection during alphavirus assembly and exit. J Virol 87(18):10255–10262. https://doi.org/10.1128/JVI.01507-13

Urakova N, Kuznetsova V, Crossman DK, Sokratian A, Guthrie DB, Kolykhalov AA et al (2018) β-D-N (4)-hydroxycytidine is a potent anti-alphavirus compound that induces a high level of mutations in the viral genome. J Virol 92(3). https://doi.org/10.1128/jvi.01965-17

Utt A, Quirin T, Saul S, Hellstrom K, Ahola T, Merits A (2016) Versatile trans-replication systems for Chikungunya virus allow functional analysis and tagging of every replicase protein. PLoS ONE 11(3):e0151616. https://doi.org/10.1371/journal.pone.0151616

van der Heijden MW, Bol JF (2002) Composition of alphavirus-like replication complexes: involvement of virus and host encoded proteins. Arch Virol 147(5):875–898. https://doi.org/10.1007/s00705-001-0773-3

van Duijl-Richter MK, Hoornweg TE, Rodenhuis-Zybert IA, Smit JM (2015) Early events in Chikungunya virus infection-from virus cell binding to membrane fusion. Viruses 7(7):3647–3674. https://doi.org/10.3390/v7072792

Vasiljeva L, Merits A, Golubtsov A, Sizemskaja V, Kaariainen L, Ahola T (2003) Regulation of the sequential processing of Semliki Forest virus replicase polyprotein. J Biol Chem 278 (43):41636–41645. https://doi.org/10.1074/jbc.M307481200

Venturi G, Di Luca M, Fortuna C, Remoli ME, Riccardo F, Severini F et al (2017) Detection of a Chikungunya outbreak in Central Italy, August–September 2017. Euro Surveill 22(39). https://doi.org/10.2807/1560-7917.es.2017.22.39.17-00646

Voss JE, Vaney MC, Duquerroy S, Vonrhein C, Girard-Blanc C, Crublet E et al (2010) Glycoprotein organization of Chikungunya virus particles revealed by X-ray crystallography. Nature 468(7324):709–712. https://doi.org/10.1038/nature09555

Wada Y, Orba Y, Sasaki M, Kobayashi S, Carr MJ, Nobori H et al (2017) Discovery of a novel antiviral agent targeting the nonstructural protein 4 (nsP4) of Chikungunya virus. Virology 505:102–112. https://doi.org/10.1016/j.virol.2017.02.014

Wang QY, Bushell S, Qing M, Xu HY, Bonavia A, Nunes S et al (2011) Inhibition of dengue virus through suppression of host pyrimidine biosynthesis. J Virol 85(13):6548–6556. https://doi.org/10.1128/JVI.02510-10

Weaver SC, Lecuit M (2015) Chikungunya virus and the global spread of a mosquito-borne disease. N Engl J Med 372(13):1231–1239. https://doi.org/10.1056/NEJMra1406035

Weber C, Berberich E, von Rhein C, Henss L, Hildt E, Schnierle BS (2017) Identification of functional determinants in the Chikungunya virus E2 protein. PLoS Negl Trop Dis 11(1): e0005318. https://doi.org/10.1371/journal.pntd.0005318

Weber C, Sliva K, von Rhein C, Kummerer BM, Schnierle BS (2015) The green tea catechin, epigallocatechin gallate inhibits Chikungunya virus infection. Antiviral Res 113:1–3. https://doi.org/10.1016/j.antiviral.2014.11.001

Weiss B, Geigenmuller-Gnirke U, Schlesinger S (1994) Interactions between sindbis virus RNAs and a 68 amino acid derivative of the viral capsid protein further defines the capsid binding site. Nucleic Acids Res 22(5):780–786. https://doi.org/10.1093/nar/22.5.780

Weston JH, Graham DA, Branson E, Rowley HM, Walker IW, Jewhurst VA et al (2005) Nucleotide sequence variation in salmonid alphaviruses from outbreaks of salmon pancreas disease and sleeping disease. Dis Aquat Organ 66(2):105–111. https://doi.org/10.3354/dao066105

Wichit S, Hamel R, Bernard E, Talignani L, Diop F, Ferraris P et al (2017) Imipramine inhibits Chikungunya virus replication in human skin fibroblasts through interference with intracellular cholesterol trafficking. Sci Rep 7(1):3145. https://doi.org/10.1038/s41598-017-03316-5

Wintachai P, Thuaud F, Basmadjian C, Roytrakul S, Ubol S, Desaubry L et al (2015) Assessment of flavaglines as potential Chikungunya virus entry inhibitors. Microbiol Immunol 59(3):129–141. https://doi.org/10.1111/1348-0421.12230

Wintachai P, Wikan N, Kuadkitkan A, Jaimipuk T, Ubol S, Pulmanausahakul R et al (2012) Identification of prohibitin as a Chikungunya virus receptor protein. J Med Virol 84(11):1757–1770. https://doi.org/10.1002/jmv.23403

Wong KZ, Chu JJH (2018) The interplay of viral and host factors in Chikungunya virus infection: targets for antiviral strategies. Viruses 10(6). https://doi.org/10.3390/v10060294

Yang S, Fink D, Hulse A, Pratt RD (2017) Regulatory considerations in development of vaccines to prevent disease caused by Chikungunya virus. Vaccine 35(37):4851–4858. https://doi.org/10.1016/j.vaccine.2017.07.065

Zhang R, Kim AS, Fox JM, Nair S, Basore K, Klimstra WB et al (2018) Mxra8 is a receptor for multiple arthritogenic alphaviruses. Nature 557(7706):570–574. https://doi.org/10.1038/s41586-018-0121-3

The manufacturer's authorised representative in the EU is Springer Nature Customer Service Centre GmbH, Europaplatz 3, 69115 Heidelberg, Germany. If you have any concerns regarding our products, please contact ProductSafety@springernature.com

Printed and bound by CPI Group (UK) Ltd, Croydon, CR0 4YY
30/09/2026
02225262-0006